Climate Shock

CLIMATE SHOCK

THE ECONOMIC CONSEQUENCES OF A HOTTER PLANET

Gernot Wagner
Martin L. Weitzman

PRINCETON UNIVERSITY PRESS
PRINCETON AND OXFORD

Copyright © 2015 by Princeton University Press

Published by Princeton University Press, 41 William Street, Princeton, New Jersey 08540
In the United Kingdom: Princeton University Press, 6 Oxford Street, Woodstock, Oxfordshire OX20 1TW
press.princeton.edu

All Rights Reserved

Library of Congress Cataloging-in-Publication Data

Wagner, Gernot, 1980–
  Climate shock : the economic consequences of a hotter planet / Gernot Wagner and Martin L. Weitzman.
      pages cm
  Includes bibliographical references and index.
  ISBN 978-0-691-15947-8 (hardcover : alk. paper) 1. Climatic changes—Economic aspects. I. Weitzman, Martin L., 1942– II. Title.
  QC903.W34 2015
  363.738'74—dc23   2014028540

British Library Cataloging-in-Publication Data is available

This book has been composed in Sabon Next LT Pro, DINPro

Printed on acid-free paper. ∞

Printed in the United States of America

1 3 5 7 9 10 8 6 4 2

To Siri and Jennifer
‖‖‖‖‖‖‖‖‖‖‖‖‖‖‖‖‖‖‖‖‖‖‖‖‖‖‖‖‖‖‖‖‖‖‖‖

Contents

IIIIIIIIIIIIIIIII

Preface

||||||||||||||

Pop Quiz

Two quick questions:

Do you think climate change is an urgent problem?

Do you think getting the world off fossil fuels is difficult?

If you answered "Yes" to both of these questions, welcome. You'll nod along, on occasion even cheer, while reading this book. You'll feel reaffirmed.

You are also in the minority. The vast majority of people answer "Yes" to one or the other question, but not both.

If you answered "Yes" *only* to the first question, you probably think of yourself as a committed environmentalist. You may think climate change is *the* issue facing society. It's bad. It's worse than most of us think. It's hitting home already, and it will strike us with full force. We should be pulling out all the stops: solar panels, bike lanes, the whole lot.

You're right, in part. Climate change is an urgent problem. But you're fooling yourself if you think getting off fossil fuels will be simple. It will be one of the most difficult challenges modern civilization has ever faced, and it will require the most sustained, well-managed, globally cooperative effort the human species has ever mounted.

If you answered "Yes" *only* to the second question, chances are you don't think climate change is the defining problem of our generation. That doesn't necessarily mean you're a "skeptic" or "denier" of the underlying scientific evidence; you may still think global warming is worthy of our attention. But realism dictates that we can't stop life as we know it to mitigate a problem that'll take decades or centuries to show its full force. Look, some people are suffering right now because of *lack* of energy. And whatever the United States, Europe, or other high emitters do to rein in their energy consumption will be nullified by China, India, and the rest catching up with the rich world's standard of living. You know there are trade-offs. You also know that solar panels and bike lanes alone won't do.

You, too, are right, but none of that makes climate change any less of a problem. The long lead time for solutions and the complex global web of players are precisely why we must act decisively, today.

||||||||||||||||||

If you are an economist, chances are you answered "Yes" to the second question. Standard economic treatments all but prescribe the stance of the "realist." After all, economists live and breathe trade-offs. Your love for your children may go beyond anything in this world, but as economists we are obligated to say that, strictly speaking, it's not infinite. As a parent, you may invest enormous sums of money and time into your children, but you, too, face trade-offs: between doing your day job and reading bedtime stories, between indulging now and teaching for later.

Trade-offs are particularly relevant on an average, national, or global level. And they are perhaps nowhere more apparent on the planetary scale than in the case of climate change. It's the ultimate battle of growth versus the environment. Stronger climate policy now implies higher, immediate economic costs. Coal-fired power plants will become obsolete sooner or won't be built in the first place.

That comes with costs, for coal plant owners and electricity consumers alike. The big trade-off question then is how these costs compare to the benefits of action, both because of lower carbon pollution and because of economic returns from investing in cleaner, leaner technologies today.

Economists often cast themselves as the rational arbiters in the middle of the debate. Our air is worse now than it was during the Stone Age, but life expectancy is a lot higher, too. Sea levels are rising, threatening hundreds of millions of lives and livelihoods, but societies have moved cities before. Getting off fossil fuels will be tough, but human ingenuity—technological change—will surely save the day once again. Life will be different, but who's to say it will be worse? Markets have given us longer lives and untold riches. Let properly guided market forces do their magic.

There's a lot to be said for that logic. But the operative words are "properly guided." What precisely are the costs of unabated climate change? What's known, what's unknown, what's unknowable? And where does what we don't know lead us?

That last question is *the* key one: Most everything we know tells us climate change is bad. Most everything we don't know tells us it's probably much worse.

"Bad" or "worse" doesn't mean hopeless. In fact, almost every prediction in this book is prefaced by a version of the words *unless we act.* We don't venture predictions only to see them become true. We talk about where unfettered economic forces may lead in order to guide them in a more productive, better direction. And guide we can. In many ways, putting a proper price on carbon isn't a question of *if*, it's a question of *when*.

Climate Shock

||||||||||||||||||||||||||||||

CHAPTER 1

911

Tʜᴀɴᴋ Rᴜssɪᴀɴ ᴘᴏʟɪᴄᴇ ᴄᴏʀʀᴜᴘᴛɪᴏɴ for footage that eluded NASA and every other space agency. On February 15, 2013, an asteroid as wide as 20 meters (66 feet) exploded in the sky above the Russian city of Chelyabinsk during the morning commute hours, causing a blast brighter than the sun. It didn't take long for some spectacular videos to appear online, mostly from dashboard cameras many Russian drivers have to protect themselves against the whims of traffic cops. The blast injured 1,500, most because of glass shattered by the explosion. It was a sobering wakeup call for space agencies to ramp up their asteroid detection and defense capabilities.

The money for such efforts is perennially in short supply. But the technical means are there, or at least they could be. A U.S. National Academy study estimates it would take ten years and around $2 or 3 billion to launch a test to deflect an asteroid bound to hit Earth. It may not be as glamorous as sending a man to the moon within the decade, but it may be at least as important.

While the Chelyabinsk asteroid would have been too small to deflect, it would have still been nice to know about it in advance. The chance of a larger asteroid hitting us is small, but it's there. Educated guesses put it as a 1-in-1,000-year event. That's a 10 percent chance each century. We haven't yet spent the money to know for sure. The fact, though, is that a few billion dollars would allow

NASA and others both to catalogue the hazards and to defend against them. That's a small amount when measured against the costs of a potentially civilization-destroying threat. Around 65 million years ago it was a giant asteroid that caused the globe's fifth major extinction event, killing the dinosaurs.

Climate change isn't exactly hurtling toward us through outer space. It's entirely homegrown. But the potential devastation is just as real. Elizabeth Kolbert argues convincingly based on her book *The Sixth Extinction* how this time around: "We are the asteroid." In fact, by one recent scientific assessment, we are slated to experience global changes at rates that are at least ten times faster than at any point in the past 65 million years.

IIIIIIIIIIIIIIIII

As Hurricane Sandy was whipping the Eastern Seaboard, leaving Manhattan below the Empire State Building partially flooded and almost entirely without power, New York governor Andrew Cuomo wryly told President Barack Obama that: "We have a 100-year flood every two years now." Hurricane Irene in August 2011 caused the first-ever preemptive weather-related shutdown of the entire, century-old New York City subway and bus system. It took only fourteen months for the second shutdown. Sandy hit in October 2012. All told, Irene killed 49 and displaced over 2.3 million. Sandy killed 147 and displaced 375,000.

New York, of course, is far from unique here. Typhoon Haiyan slammed the Philippines in November 2013, killing at least 6,000 people and displacing four million. Not even a year earlier, Typhoon Bopha struck the country, killing over a thousand and displacing 1.8 million. The

European summer heat wave in 2003 killed 15,000 in France alone, over 70,000 in Europe. The list goes on, spanning both poor and rich countries and continents.

Society as a whole—especially in rich places like the United States and Europe—has never been as well equipped to cope with these catastrophes as it is today. As is so often the case, the poor suffer the most. That makes these recent deaths and displacements in places like New York all the more remarkable.

What likens these storms and other extreme climatic events to asteroids is that they both can be costly, in dollars and in deaths. The important and clear differences show that the climate problem is costlier still.

First the obvious: Major storms have hit long before humans started adding carbon dioxide to the atmosphere. However, warmer average temperatures imply more energy in the atmosphere implies more extreme storms, floods, and droughts. The waters off the coast of New York were 3°C (5.4°F) warmer than average during the days before Sandy. The waters off the coast of the Philippines were 3°C (5.4°F) warmer than average just as Haiyan was intensifying on its path to make landfall. Coincidence? Perhaps. The increase off New York happened at the surface. The increase off the Philippines happened 100 meters (330 feet) below. But the burden of proof seems to rest on those questioning the link from higher temperatures to more intense storms.

That's particularly true, since the best research goes much beyond drawing circumstantial links. The science isn't settled yet, but the latest research suggests that climate change will lead both to more *and* bigger storms. Though hurricanes are among the toughest climatic events to link directly to climate change, mainly because of how rare they are. It's easier to draw the direct link from climate change to more common events like extreme temperatures, floods, and droughts.

Think of it like drunk driving: Drinking increases the chance of a car crash, but plenty of crashes happen without elevated blood alcohol levels. Or liken it to doping in sports: No single Barry Bonds home run or Lance Armstrong Tour de France stage win can be attributed to doping, nor did doping act alone. Bonds still had to hit the ball, and Armstrong still had to pedal. But doping surely helped them hit farther and bike faster. Major storms, like home run records and multiple *Le Tour* wins, have happened before. None of that means steroids or elevated levels of red blood cells in an athlete's blood had no effect. Something similar holds for elevated levels of carbon dioxide in the atmosphere.

Researchers are getting increasingly better at using "attribution science" to identify the human footprint even in single events. The UK's National Weather Service, more commonly known as the Met Office, has a Climate Monitoring and Attribution team churning out studies that do just that. One such study found with 90 percent confidence that "human influence has at least doubled the risk of a heatwave exceeding [a] threshold magnitude" of mean summer temperature that was met in Europe in 2003, and in no other year since 1851. Links will only become clearer in the future, both because the science is getting better and because extreme weather events are becoming ever more extreme.

Governor Cuomo's "100-year flood every two years" comment may have been a throw-away line, but he was on to something. By the end of the century, we can expect today's 100-year flood to hit as frequently as once every three to twenty years. That's a century out, long after our lifetimes, but we know that we can't wait that long to act. Already, the annual chance of storm waters breaching Manhattan seawalls has increased from around 1 percent in the 19th century to 20 to 25 percent today. That means lower Manhattan can expect some amount of flooding every four to five years.

Unlike with asteroids, there's no $2-to-3-billion, ten-year NASA program to avoid the impact of storms and other extreme climatic events like floods and droughts. Nor is there a quick fix for less dramatic events like the ever faster rising seas. As a first line of defense, higher seawalls would surely help. But they can go only so far for so long. Higher seas make storm surges all the more powerful, and higher seas themselves come with plenty of costs of their own. Imagine standing in the harbor of your favorite coastal city. Then imagine standing there at the end of the century with sea levels having risen by 0.3 to 1 meters (1 to 3 feet). It will only be a matter of time before higher seawalls won't do, when the only option will be retreat.

By then, it will be too late to act. We can't re-create glaciers and polar ice caps, at least not in human timescales. The severity of the problems will have been locked in by past action, or lack thereof. Future generations will be largely powerless against their own fate.

One possible response that attempts to provide a quick fix is large-scale geoengineering: shooting small reflective particles into the stratosphere in an attempt to cool the planet. Geoengineering is far from perfect. It comes with lots of potential side effects, and it's no replacement for decreasing emissions in the first place. Still, it may be a useful, temporary complement to more fundamental measures.

(We will start exploring the full implications of geoengineering in chapter 5.)

||||||||||||||||

None of what we've talked about thus far even deals with the true worst-case scenarios. Having the climatic equivalent of ever more Chelyabinsk-like asteroids hit us is bad, but there are ways to cope. For relatively small asteroids, it's seeking shelter and moving away from windows. For relatively small climatic changes, it's moving to slightly cooler climates and higher shores. That's often easier said than done, but at least it's doable. For much more dramatic climatic consequences—such as a crippling of the world's productive agricultural lands—it's tough to imagine how we'd cope in a way that wouldn't cause serious disruptions.

Meanwhile, standard economic models don't include much of this thinking. Many observers regard average global warming of greater than 2°C (3.6°F) above preindustrial levels as having the potential to trigger events deserving of various shades of the label "catastrophe." Economists typically have a hard time making sense of that term. They need dollar figures. Does a catastrophe then cost 10 percent of global economic output? 50 percent? More?

While it's indeed necessary to translate impacts into dollars and cents, such benefit-cost analyses can act as only one guide for how society ought to respond. We should also take into account the potential for planet-as-we-know-it-altering changes in the first place. First and foremost, climate change is a risk management problem—a catastrophic risk management problem on a planetary scale, to be more precise.

CAMELS IN CANADA

If one wanted to imagine an all but intractable public policy problem, climate change would be pretty close to the ideal. Today's storms, floods, and wildfires notwithstanding, the worst effects of global warming will be felt long after our lifetimes, likely in the most unpredictable of ways. Climate change is unlike any other environmental problem, really unlike any other public policy problem. It's almost uniquely *global*, uniquely *long-term*, uniquely *irreversible*, and uniquely *uncertain*—certainly unique in the combination of all four.

These four factors, call them the *Big Four*, are what make climate change so difficult to solve. So difficult that—short of a major jolt of the global, collective conscience—it may well prove too difficult to tackle climate change just by decreasing emissions and adjusting to some of the already unavoidable consequences. At the very least we'll need to add *suffering* to the list. The rich will adapt. The poor will suffer.

Then there's the almost inevitable-sounding geoengineering, attempting a global-scale techno fix for a seemingly intractable problem. The most prominent geoengineering idea would have us deliver tiny sulfur-based particles into the stratosphere in an attempt to engineer an artificial sun shield of sorts to help cool the planet.

Everything we know about the economics of climate change seems to point us in that direction. Geoengineering is so cheap to do crudely, and it has such high leverage, that it almost has the exact opposite properties of carbon pollution. It's the "free-rider" effect of carbon pollution that has caused the problem: it's in no one's narrow self-interest to do enough. It's the "free-driver" effect that may push us to geoengineer our way out of it: it's so cheap that someone will surely do it based on their own self-interest, broader consequences be damned.

But let's not go there quite yet. Let's first tackle the Big Four in turn, beginning with why climate change is the ultimate "free-rider" problem:

Climate change is uniquely global. Beijing's smog is bad. So bad, that it comes with real and dramatic health effects that have prompted city officials to close schools and take other drastic actions. But Beijing's smog—or that in Mexico City or Los Angeles, for that matter—is mostly confined to the city. Chinese soot may register at measuring stations on the U.S. West Coast, much like Saharan dust may on occasion blow to central Europe. But all these effects are still regional.

That's not true for carbon dioxide. It doesn't matter where on the planet a ton is being emitted. Impacts may be regional, but the phenomenon is global and—among environmental problems—almost uniquely so. The ozone hole over the Antarctic is bad, but even at its height it has never reached the level of engulfing the globe. The same goes, say, for biodiversity loss or deforestation. These are regional problems. It's climate change that ties them together into phenomena with global implications.

The global nature of global warming is also strike one against enacting sensible climate policy. It's tough enough to get voters to enact pollution limits on themselves, when those limits benefit them and only them, and when the benefits of action outweigh the costs. It's a whole lot tougher to get voters to enact pollution limits on themselves if the costs are felt domestically but the benefits are global: a planetary "free-rider" problem.

Climate change is uniquely long-term. The past decade was the warmest in human history. The one before was the second-warmest. The one before that was the third-warmest. "Americans are noticing changes all around them," as the 2014 U.S. National Climate Assessment puts it. Changes are nowhere as evident as above the Arctic Circle: Arctic sea ice has lost half of its area and three-quarters of its volume in only the past thirty years. The *Foreign Policy* article describing "The Coming Arctic Boom" takes all of this as given. Then there are the visible changes all around. Again, from the National Climate Assessment: "Residents of some coastal cities see their streets flood more regularly during storms and high tides. Inland cities near large rivers also experience more flooding, especially in the Midwest and Northeast. Insurance rates are rising in some vulnerable locations, and insurance is no longer available in others. Hotter and drier weather and earlier snowmelt mean that wildfires in the West start earlier in the spring, last later into the fall, and burn more acreage." Climate change is here, and it's here to stay.

None of that should mask the fact that most of the worst consequences of climate change are still remote, often caged in global, long-term averages: global average surface temperature projections for 2100, or global average sea level projections for decades and centuries out. Strike two against sensible climate policy: the worst effects are far off—never mind that avoiding these predictions would entail acting now.

Climate change is uniquely irreversible. Even if we stopped emitting carbon tomorrow, we would have decades of warming and centuries of sea-level rise locked in. The eventual, full melting of large West Antarctic ice sheets may already be unstoppable. More extreme weather events are already here and will be with us for some time to come.

Over two-thirds of the excess carbon dioxide in the atmosphere that wasn't there when humans started burning coal will still be present a hundred years from now. Well over one-third will still be there in 1,000 years. These changes are long-term, and—at least in human timescales—virtually irreversible. Strike three.

||||||||||||||||

As if three strikes weren't enough, there's another unique characteristic of climate change to round out the Big Four, and it may be the biggest one of them all: *uncertainty*—everything we know that we don't know, and perhaps more importantly, what we don't yet know we don't know.

Last time concentrations of carbon dioxide were as high as they are today, at 400 parts per million (ppm), the geological clock read "Pliocene." That was over three million years ago, when natural variations, not cars and factories, were responsible for the extra carbon in the air. Global average temperatures were around 1–2.5°C (1.8 to 4.5°F) warmer than today, sea levels were up to 20 meters (66 feet) higher, and camels lived in Canada.

We wouldn't expect any of these dramatic changes *today*. The greenhouse effect needs decades to centuries to come into full force. Despite the recent changes in the Arctic, ice sheets need decades to centuries to melt. Global sea levels take decades to centuries to adjust accordingly. Carbon dioxide concentrations may have been at 400 ppm three million years ago, whereas rising sea levels lagged decades or centuries behind. That time difference is important and points to the long-term nature and irreversibility of it all. See strikes two and three.

But all that's small consolation, and there's an important twist to strike four.

DEEP UNCERTAINTIES

The best available climate models come close in their temperature projections to what the world experienced during the Pliocene, but they aren't predicting sea levels of 20 meters (66 feet) higher. Nor do they predict camels wandering around Canada. Not now. Not hundreds of years from now. That's true for two important reasons.

First, most climate models are unduly skewed toward the known, sometimes making them much too conservative. Until recently, most climate models predicted rising sea levels only based on thermal expansion of the oceans (and the melting of mountain glaciers), but they did not include the effects of melting ice sheets. Warmer waters take up more space, leading to higher sea levels. That mechanism alone has indeed contributed to over a third of sea-level rise in the past two decades. It's also clear that melting glaciers in Greenland and Antarctica raise sea levels, but by how much is highly uncertain. Call it a "known unknown." Until recently, scientific understanding of melting polar ice caps had been so poor that most models simply left it out.

Second, even though climate models do get a lot of things right, there are fundamental things that we don't understand about the way the climate works. The averages are bad enough. While 0.1°C (0.2°F) of average global surface warming per decade sounds rather manageable and perhaps even pleasant, few dispute that a century or more of warming at this pace would lead to serious costs. But these averages hide two distinct sets of uncertainties that could pose the real problems.

The first set of uncertainties is inherent in any kind of global, long-term estimate. Presenting just the global average numbers masks at least four important facts: First, temperatures in the past century have been increasing at an increasing rate. Second, despite that generally increasing trend, temperatures fluctuate across years and decades. (Hence the infamous "decade without warming.") Third, air over the oceans is usually cooler than over land. Since two-thirds of the world is ocean, a global average increase of 0.07°C (0.13°F) per decade translated to about a 0.11°C (0.20°F) increase over land. Finally, temperatures over the poles have warmed more than elsewhere. Arctic temperatures are expected to increase at a rate more than twice the global average. That's particularly bad, since the poles are also where most of the world's remaining ice is. Melting ice on land above sea level means higher seas, as the latest sea-level projections now officially acknowledge.

Then there are the real, deep-seated uncertainties. To arrive at any of these projections—average or otherwise—requires taking several steps, each with its own set of known and, most vexingly, unknown unknowns. Uncertainties exist around the amounts of global warming pollutants we emit, the link between emissions and atmospheric concentrations, the link between concentrations and temperatures, the link between temperatures and physical climate damages, the link between physical damages and their consequences, and, at least as important, how society will respond: what coping measures will be undertaken, and how effective they will prove to be.

Nailing down one of these steps—the link between concentrations and eventual temperature increases—has proven particularly elusive. The past three decades of amazing advances in climate science have gotten us no closer to pinpointing the true answer. Double the carbon dioxide concentrations in the atmosphere—something that will surely happen, unless we enact ambitious climate policies now—and eventually global average temperatures are *likely* to go up by between 1.5 and 4.5°C (2.7 and 8°F). Our confidence in that range has increased, but what's now called the "likely" range hasn't changed since the late 1970s, a fact we will revisit in chapter 3, "Fat Tails."

The very term "fat tails" also points to another problem: 1.5 to 4.5°C (2.7 to 8°F) is "likely" in the best sense of that word. The chance is good that we will indeed find ourselves somewhere in that range for how temperatures react when concentrations double, what's known as "climate sensitivity." But there's also a chance we won't. The Intergovernmental Panel on Climate Change (IPCC) describes anything below 1°C (1.8°F) as "extremely unlikely." That assessment is pretty believable, given that the world has already warmed by 0.8°C (1.4°F), and we haven't even yet doubled carbon dioxide concentrations from preindustrial levels. (The 400 ppm that the world just passed is a 40 percent increase over preindustrial levels of 280 ppm.) There's also a chance that final temperatures from a doubling of carbon dioxide concentrations will end up above 4.5°C (8°F). It's "unlikely," but we can't discount the possibility.

Meanwhile, global average warming of 4.5°C (8°F) is beyond the pale of most imagination. Recall the camels in Canada, or at least a planet that none of us would recognize.

But that 4.5°C (8°F) doesn't yet tell the full story. Climate sensitivity describes what happens when concentrations of carbon dioxide in the atmosphere double. What if carbon dioxide concentrations more than double? The International Energy Agency (IEA) predicts levels of 700 ppm, or two-and-a-half times preindustrial levels. Now we are looking at a "likely" range of temperatures between 2 and 6°C (3.6 and 11°F).

Climate science warns that average global warming above 2°C (3.6°F) could trigger potentially devastating events. It's unclear what label to use for global average warming of 6°C (11°F): "catastrophic" no longer seems to do it justice. Mark Lynas, who has painstakingly detailed climate impacts degree by frightening degree, ends his book *Six Degrees* just there. The introduction to the final chapter on 6°C (11°F) begins with a reference to Dante's Sixth Circle of Hell. HELIX, a recently started project funded by the European Union, aims to determine global and regional impacts of specific levels of temperature rise. It, too, ends at 6°C (11°F). And per our own calculations in chapter 3, we are looking at an eventual chance of around 10 percent of *exceeding* that mark.

|||||||||||||||||||||

Whenever science points to the very real potential of these types of catastrophic outcomes, cognitive dissonance kicks in. Facts might be facts, the reasoning goes, but throwing too many of them at you at once will all but guarantee that you will dismiss them out of hand. It just *feels* like it can't or shouldn't be true.

That fickleness of human nature and the limits of our understanding are at the core of the climate policy dilemma. Smarts alone don't seem to make much of a difference here. Solving the dilemma will take a completely different way of thinking.

THE BATHTUB PROBLEM

Think of the atmosphere as a giant bathtub. There's a faucet—emissions from human activity—and a drain—the planet's ability to absorb that pollution. For most of human civilization and hundreds of thousands of years before, the inflow and the outflow were in relative balance. Then humans started burning coal and turned on the faucet far beyond what the drain could handle. The levels of carbon in the atmosphere began to rise to levels last seen in the Pliocene, over three million years ago.

What to do? That's the question John Sterman, an MIT professor, asked two hundred graduate students. More specifically, he asked what to do to stabilize concentrations of carbon dioxide in the atmosphere close to present levels. How far do we need to go in turning off the faucet in order to stabilize concentrations?

Here's what not to do: stabilizing the flow of carbon into the atmosphere today won't stabilize the carbon already there at close to present levels. You're still adding carbon. Just because the inflow remains steady year after year, doesn't mean the amount already in the tub doesn't go up. Inflow and outflow need to be in balance, and that won't happen at current levels of carbon dioxide in the tub (currently at 400 ppm) unless the inflow goes down by a lot.

That seems like an obvious point. It also seems to get lost on the average MIT graduate student, and these students aren't exactly "average." Still, over 80 percent of them in Sterman's study seem to confuse the faucet with the tub. They confuse stabilizing the inflow with stabilizing the level.

To be fair, these two hundred MIT students weren't told about the bathtub analogy. They just saw an excerpt of the "Summary for Policymakers" from the latest IPCC report at the time. That's the document that's meant to explain the issue to our elected officials. If as fundamental a point as the difference between annual emissions and concentrations in the atmosphere—the difference between the inflow and the level of carbon in the tub—is lost on MIT graduate students, what hope is there for the rest of us?

Sure, it's a "Summary for *Policymakers*." Jane Q. Public may not need to understand it, as long as policy makers do. But there, too, is a hiccup. MIT graduate students may well be a good proxy for (better-educated) policy makers. Moreover, there are policy makers, and there are policy makers. The anonymous bureaucrat writing the actual policies may have a Ph.D. in the subject for which he or she is making policy. One hopes. The elected official is unlikely to be a specialist in any particular subject. And ultimately, of course, Jane Q. Voter decides how that person ought to think about a particular issue.

It shouldn't come as a surprise then that one all too popular option among elected officials is a so-called wait-and-see approach to tackling global warming pollution. It's precisely what it sounds like, and it's as misguided as the bathtub analogy would suggest. We can't wait until the moment when that crucial Antarctic ice sheet slips into the ocean and brings us 3 meters (10 feet) closer to where global sea levels were in the Pliocene. At that point, even the last holdouts would realize we are in a climatic emergency. But the emergency is linked to the concentration of carbon in the atmosphere. Society can most directly control the inflow of emissions, and even turning that inflow to zero immediately wouldn't solve the problem. It will take centuries and millennia for the excess carbon to flow out naturally. "Wait and see" might as well be called "give up and fold."

|||||||||||||||||||

Climate change requires an entirely new way of thinking, something as seemingly foreign to MIT graduate students as to policy makers and the general public. And lest we think getting serious about climate change is as simple as understanding the bathtub analogy and acting accordingly—as seemingly difficult as that alone is—this analogy highlights only two of the Big Four issues: the long-term nature of climate change, with a whole lot of irreversibility mixed in. Nothing yet on the other two: how global and uncertain climate change truly is. The global nature of global warming all but guarantees that deliberately turning off the faucet is incredibly tough to do. Uncertainty doesn't exactly help either, even though it ought to prompt stronger action today. If you don't know precisely how far the tub is from flooding, it's only prudent to turn off the faucet sooner.

WE CAN DO THIS

There are plenty of angles to take from here.

One can try to be optimistic. Yes, things are dire, but look at all the progress. The price of solar panels has declined by 80 percent within five years. Much of it has happened on the backs of German and Chinese households, whose governments took to direct subsidies to bring down costs, but the best way to respond may be to brush up on your German and Chinese for those thank-you notes. They took the hit, for the rest of us to enjoy cheaper solar energy.

Solar energy is not a perfect replacement for fossil sources, at least without significant improvements to electricity market structures and storage technologies. A coal or gas plant can be turned on and off, but we can't control when the sun shines. Still, on a sunny Sunday afternoon, when the sun is up and demand is down, Germany gets 50 percent of its electricity from the sun. Averaged over the entire year of 2013, Germany got almost 5 percent of its electricity from the sun. That's Germany, the industrial powerhouse in Europe, not typically thought of as a particularly sunny place.

Things are looking up globally, too. The world added almost 40 gigawatts (GW) of total solar capacity in 2013, on top of the 30 GW added in 2012, which came on top of the 30 GW added in 2011. The absolute numbers are large, but the rate of change is even more significant. In 2000, the world had around 1 GW of total installed solar capacity. At the end of 2010, the world had 40 GW. By the end of 2013, the tally stood at 140 GW. That's explosive growth on overdrive.

And the all-important policy changes are happening as we speak. None yet is sufficient in itself, but together they provide an impressive array of policy frameworks. Europe has had its carbon market up and (fully) running since 2008. By now, California has the world's most comprehensive carbon market, covering 80 percent of its total greenhouse gas emissions. British Columbia has a carbon tax. China is experimenting with seven regional carbon market trials, and it has a commitment to peak its carbon emissions by 2030. India has a $1-per-ton coal tax. Not a lot, but it's there, and it's positive. Brazil has an ambitious national climate target and has sharply reduced carbon emissions from deforestation. And—since we're being optimistic—in the United States, a solid majority of the electorate would like elected officials to act, at least in principle. A handful more 100-year storms like the two that hit New York City within the course of two years in 2011 and 2012, and we may well see real change.

In fact, the path toward sensible U.S. climate policy is becoming increasingly clear. For one, it will likely go via state capitals like Sacramento. It will also be going through the Clean Air Act and the Environmental Protection Agency's carbon pollution standards for new and existing power plants. At the very least, these regulations could provide a real bargaining chip when it comes to U.S. Congress considering comprehensive climate policy and a direct price on carbon down the line.

IIIIIIIIIIIIIIIII

Optimism is good. Economics as a discipline is almost pathologically optimistic, even though it's often seen to be a different kind of optimism. Growth is good. Trade is good. Technology is good. There are asterisks for every one of these statements, but they are just that. Few economists may believe that solar panels will save the day, but new technologies have pulled us out of deep environmental morasses in the past—quite literally. New technologies solved the horse manure crisis threatening to engulf New York City at the end of the 19th century. The internal combustion engine banished horses and buggies to taking tourists around Central Park. No one predicted that particular invention at the time. And it didn't require much in terms of active policy intervention: invent car + find oil = *Eureka!*

There may well be one of these breakthroughs just around the corner. Human history seemingly shows that there always is. It's why we are still here as a species. But hoping for a breakthrough is not a strategy. That's why we return to the undeniable importance of policy. That, too, has worked in the past.

For many pollutants, things first got (and are getting) worse, before they got (or will get) better. When Cleveland's Cuyahoga River caught on fire, so did the nascent environmental movement in the United States in the 1960s. This, in turn, led Richard Nixon to sign into law the National Environmental Policy Act of 1969 and create the Environmental Protection Agency. And that was just the beginning. In addition, Nixon went on to sign the Clean Air Act of 1970, the Clean Water Act in 1972, and the Endangered Species Act in 1973, to name just the major ones. A dozen more laws helped round out the "environmental decade." And the U.S. Congress has acted boldly since, with large bipartisan majorities. George H. W. Bush signed into law the Clean Air Act Amendments of 1990. Among others, they led to measures that slashed the pollution that causes acid rain.

All of that applies to local pollutants: the mercury knocking a few points off your kids' IQ, the soot causing them to develop early asthma, the smog making their eyes water and killing their grandparents early, and the toxins in water making it unsafe for anyone to drink. You see, smell, or feel the problem. You petition your government. It reacts. Problem solved.

In reality, it is, of course, much messier than this simple chain would suggest. Niccolò Machiavelli put it succinctly in *The Prince*, published in 1532: "There is nothing more difficult to take in hand, more perilous to conduct, or more uncertain in its success, than to take the lead in the introduction of a new order of things. Because the innovator has for enemies all those who have done well under the old conditions, and lukewarm defenders in those who may do well under the new."

London experienced its first major bout with air pollution in the 1280s. King Edward I established the first air pollution commission in 1285. In 1306, he made it illegal to burn coal. The punishment for repeat offenders: death. You'd think that with the right amount of monitoring and enforcement, this should have taken care of the problem. Alas, the law was soon vacated—and coal-burning has continued ever since.

Never mind all that messiness. Assume for argument's sake that addressing conventional pollutants is as easy as "see something, say something," before watching the rule of law wield its gavel. Climate change just isn't anything like local air pollution. It is, after all, more global, long-term, irreversible, and uncertain than any other environmental problem. The usual politics don't apply. For one, we don't all even agree on the problem. Reverend Martin Luther King Jr. had his dream when the nightmare was clear to most everyone at the time. We don't seem to be quite there yet on the climate front, at least not in the United States.

NO WE CAN'T

Everything we know about the basic chemistry and physics of how our atmosphere works, and everything we know about the economics of how people behave and the messy politics of how we govern ourselves, leads us to believe that things will get worse before they get better. The fact that pumping carbon dioxide into the atmosphere traps heat—the greenhouse effect—had been discovered by 1824, shown in a lab by 1859, and quantified by 1896.

By now, humans have accumulated around 940 *billion* tons of carbon dioxide in the atmosphere, and counting, enough for atmospheric concentrations of carbon dioxide to have busted through the 400 ppm mark. Concentrations are still increasing at a rate of 2 ppm a year, and that annual increase itself is still increasing.

Then there's the biggest problem, and once again a rather unique one: That continued march in the wrong direction is due to seven billion of us, or at least the billion or so high-emitters most responsible for the total number. The responsibility rests with everyone and no one. There's no finger to point. The enemy is us, all of us. The politics are messy. It's often tough to be optimistic.

For every positive piece of climate policy news, there seems to be an opposing negative one. Yes, India has a $1-per-ton coal tax. It also has about $45 billion in annual fossil fuel subsidies. China may have seven regional cap-and-trade trials. It, in turn, subsidizes fossil fuel to the tune of $20 billion annually. The world subsidizes fossil fuels at a rate of over $500 billion per year. That is equivalent to an average worldwide subsidy of some $15 per ton of carbon dioxide emissions, with lower subsidies in most developed economies and much higher per-ton subsidies in oil-rich countries like Venezuela, Saudi Arabia, and Nigeria. Every one of these dollars is a step backward for the climate. Far from moving toward the right incentives, we seem to be guiding markets in exactly the wrong direction.

||||||||||||||||

Another reason we don't always take the optimistic path is that from the economic perspective, it's rather well-trodden. We've known what needs to be done for a long time. For one, stop subsidizing fossil fuels. Now. It will be tough to make the politics work. Just ask Nigerian president Goodluck Jonathan, who stopped fuel subsidies in January 2012 and quickly backtracked, at least partially, after nationwide riots. That still doesn't make the policy prescription any less appropriate economically.

Far beyond just stopping fossil fuel subsidies, the overall policy framework needed for addressing climate change is clear and has been for decades.

THE SOLUTION TO CLIMATE CHANGE

No one is going to win the Nobel Prize in economics for finding the solution to climate change. The economist who came up with it died a decade before the first prize was given out, and the Swedes no longer award their prizes posthumously. Arthur C. Pigou identified the general problem and the solution—what's by now known as "Pigouvian taxes." Each of the 35 billion tons of carbon dioxide emitted *this year* causes at least about $40 worth of damages to the planet, possibly much more. The correct—the only correct—approach is to price each and every ton of carbon according to the damage it causes.

The average American emits about 20 tons per year. That's 20 times $40 or at least about $800 per person per year. But no one is suggesting that every American send in an $800 check at the end of the year. In fact, the entire point is not to. Every one of us ought to face the right incentives each time we turn on the heat or the air conditioner or fill up our tank of gas. At $40 per ton of carbon dioxide, that means about 35 cents per gallon of gasoline.

Pigou's crucial insight was that we ought to see and pay these costs right then and there at the pump. That's the only way to create the right incentives and lead us to incorporate the full cost into our daily decisions—and stop privatizing benefits while socializing costs.

The result of such a price on carbon dioxide will be that we use less coal, oil, and natural gas. We'd pollute less. More specifically, with the correct price we'd be polluting the "optimal" amount. That's not necessarily zero. It's certainly much less than where we are now, with one's weight in pollution going into the atmosphere every day and a half for the average American.

That's the policy solution in a nutshell: put an appropriate price on burning carbon that reflects its true cost to society.

You can get there either through a tax or by creating an explicit market for carbon dioxide emissions: cap overall emissions, allocate allowances to major emitters, and let them trade these allowances to establish a market price for pollution—"cap and trade." In a theoretical vacuum without uncertainty, the two approaches yield the exact same result. Economists love to have epic debates about which is the better approach in practice.

Taxes are simpler, one line of reasoning goes. No, they aren't. Look at the thousands of pages of the U.S. tax code.

Taxes get the price of pollution up. That's what we need. Yes, for now. But cap and trade limits emissions. That's the ultimate point. If emissions go down cheaply, all the better.

Taxes provide price certainty. Maybe, assuming no political tampering. But first off, any cap-and-trade system can be designed with price certainty in mind. It's as straightforward as creating a price floor and provisions to prevent prices from going above a certain level. And more importantly, even without any of these design features, cap-and-trade prices tend to vary just the right way: low prices during a recession, when demand for emissions allowances is low. Higher prices when business investment is strong, all the while ensuring that overall emissions decline in line with the cap.

But if cap-and-trade prices go through the roof, or collapse to zero, the entire system gets discredited. Electricity price spikes may have derailed market deregulation for generations. Sure, but we aren't talking about price spikes here. If anything, we'd expect prices that are much lower than expected because industry tends to have ways to innovate its way to lower compliance costs than previously assumed.

Taxes allow for other measures like Corporate Average Fuel Economy (CAFE) standards to show their effect. Under a cap, these types of overlapping regulations may only shift emissions but not actually reduce them. Fair enough. But that only shows the importance of getting a cap in the first place. With one, fewer of these other measures would be necessary.

That's where the debate stands at the moment, though the final chapter has yet to be written. The latest theoretical insights point to how taxes may allow for easier international coordination. In theory at least, negotiating a uniform tax rate, the proceeds from which are retained by each country, engenders an ever-so-subtle way of countervailing the force of the free-rider problem altogether. If we all agreed on a uniform tax rate per unit of carbon dioxide, then raising the tax would hurt me directly by raising my cost of using carbon-emitting energy but would help me because it makes everyone else cut down on their carbon dioxide emissions as well. By contrast, negotiating caps alone creates the clear incentive for wanting laxer caps. Negotiating a uniform, global tax can achieve something close to the global optimal outcome. That, of course, says nothing yet about the politics, which once again are the biggest hurdle.

||||||||||||||||||||

For now, just remember that, in theory *and* practice, both taxes and cap-and-trade systems implement Pigou's vision that polluters pay when they are doing the polluting and, hence, will pollute less. We, together with most economists, would be fine with either carbon taxes or caps, done correctly.

Now we can have endless discussions about how to get there in real life. How did the Swedes manage to pass the world's first tax on carbon dioxide in 1991? Why did the French fail in their efforts to enact one in 2009? Why did Europe have the world's first major carbon cap-and-trade system? What's taking the United States so long? And why are we still subsidizing fossil fuels to the tune of $15 per ton of carbon dioxide globally, when the right number should be at least $40 per ton going the other direction?

Plenty of disciplines have useful things to say about each of these questions. Political scientists, psychologists, sociologists, and climate science communicators all have their own variations of the crucial question: If—since—science has been telling us that this is such a grave problem, why hasn't the world acted accordingly?

For one, it's incredibly hard to overcome the huge vested interests fighting against Pigou's and most every economist's vision of the ideal world. Simply saying it ought to be doesn't make it so. Instead of shouting "carbon tax" or "carbon cap," economists ought to work constructively with what we have: second-, third-, and fourth-best solutions (and worse) that create all sorts of inefficiencies, unintended consequences, and other problems, but that roll with the punches of a highly imperfect policy world—and may even remove some existing, imperfect policy barriers at the same time.

Electricity grid reform is a good example. Far from sending proper signals to households and businesses, electricity prices get averaged, subsidized, and artificially stabilized for all sorts of reasons—sending distorted price signals all across the grid. Getting a price on carbon would be great, but grid reform is an essential step toward creating a level playing field for energy efficiency, demand response, and renewable energy. It's also a battle that can and needs to be fought entirely outside the U.S. Congress. It's often up to states to set policies. That alone doesn't mean the policy debate will be any more sensible—especially given how much is at stake for traditional, largely fossil-fueled utilities—but it does mean economists ought to engage much more deeply than the standard Pigouvian line about proper carbon pricing.

Gasoline prices paid at the pump are another realm where this discussion between first-best policies and reality plays out in real time. Most every economist's ideal solution to underpriced pollution from driving is to raise the price of gasoline at the pump. But instead of increasing the federal gasoline tax in the United States from 18.4 cents per gallon, its level since 1993, to something closer to the optimal level, the regulatory instrument of choice has been raising corporate average fuel-economy or CAFE standards for cars and trucks. Tightened CAFE standards were likely the one new rule from President Obama's first term with the single biggest climate impact. Opinions differ on how cost-effective CAFE standards are. What's clear is that raising CAFE standards is possible to do, even though increasing the gas tax is theoretically the first-best policy solution. Once again, it would behoove economists to engage in CAFE policy debates much beyond the level of shouting "gas tax" every chance they get.

We won't engage in either exercise. We won't be repeating the "gas tax," "carbon tax," or "carbon cap" mantras every chance we get. We also won't engage in the messy world of electricity grid reform, CAFE standards, and other policy measures that are very much necessary and also require sensible economic thinking.

TOUGHER THAN ANYTHING THAT HAS COME BEFORE

Instead, we'll go back to basic economics and zero in on two topics that move us far beyond the standard debates. In particular, we'll focus on the economics of uncertainty and geoengineering, two topics that are highly uncomfortable, highly charged, and central to understanding why climate change matters to all of us. They also show clearly why we must act now.

Climate change harbors some deep uncertainties, sometimes to the point of actual ignorance. Why don't climate models predict up to 20 meters (66 feet) of sea-level rise and camels in Canada as a result of carbon dioxide concentrations at levels from three million years ago, when the world experienced both? In short, we don't know. But uncertainty is no excuse for inaction. It's a call to tackle the climate problem while we still can.

This is a hellishly difficult problem to solve. And if the world doesn't solve it, it will hit us with full force in unpleasant and unexpected ways. This is where we'll end up:

with the specter of geoengineering. Everything we know about how humans behave, and how they don't, leads us to believe that—unless political leaders muster the courage to act, decisively and soon—the world will inevitably be facing some painful choices. It may be folly to believe that technology (in the form of geoengineering) can, once again, bail out society and the planet from the worst of planetary emergencies. But that's the world we are moving toward.

Talk of geoengineering, much like uncertainty, isn't very comforting. It shouldn't be. It's certainly not an excuse for inaction on sensible climate policy, just as we shouldn't start smoking because an experimental lung cancer drug treatment showed some promise in a lab. The specter of geoengineering should be a clarion call for action. Decisive, and soon.

We will come back to the economics of uncertainties—fat tails—and geoengineering in due course. First, a quick 411 of the other key economic concepts and the general state of the debate that will guide our journey into the unknown, unknowable, and sometimes just plain scary.

CHAPTER 2

411

BATHTUB

(1) Tub that holds water, and typically comes with a faucet and a drain.

(2) Overused analogy in which water is compared to carbon dioxide and other greenhouse gases; tries to make one of the more complex problems in all of science seem like a daily cleansing ritual. Still, climate scientists—and the rest of us—would be well advised to remind ourselves daily of its significance. Climate policy is about getting levels in the tub down.

That's extremely difficult with a tub as large as the global atmosphere, when no one person controls either the inflow or the outflow. It's the actions of seven billion people that define the human-caused inflow, and it's largely the actions of nature that define the outflow. Even there, of course, humans have an influence: deforestation, for example, clogs the drain; planting trees unclogs it.

There are natural seasonal fluctuations: more land and, thus, more vegetation in the northern hemisphere implies that carbon dioxide levels decrease during the northern growing season and increase again when significant amounts of vegetation decay during the northern hemisphere's late fall and winter. Without human interference, inflow and outflow would be roughly in balance over the course of any given year. That hasn't been the case ever since the industrial revolution.

The natural seasonal variation in global carbon dioxide levels is roughly 5 parts per million (ppm) in any given year. The current rate of increase is over 2 ppm per year. Three years' worth of fossil fuel emissions top global seasonal variations. And that annual increase of 2 ppm itself is still increasing. That fact alone makes the bathtub analogy so important.

Recall the experiment with MIT students from chapter one: it's not enough to stabilize emissions and avoid annual increases in how much carbon dioxide the world pumps into the atmosphere. We need to decrease emissions to near zero to begin to bring down concentrations. That's where the shock part of *Climate Shock* comes in. It's proven tough enough to turn the corner on carbon dioxide flows into the atmosphere, let alone turning the corner on excess carbon dioxide stocks already there.

The International Energy Agency (IEA) estimates that, without a significant course correction, the world is currently on track to increase total greenhouse gas concentrations to around 700 ppm by 2100, and levels are only going to go up further thereafter on this trajectory.

The IEA calls this trajectory the "New Policies Scenario," which takes at face value various pledges by governments to reduce emissions and assumes "continued support for renewables and efficiency, an expansion of carbon pricing and a partial removal of fossil fuel subsidies." By comparison, the world has just passed 400 ppm for carbon dioxide, somewhere between 440 and 480 ppm when counting all other greenhouse gases measured under the Kyoto Protocol. Unless we see a dramatic turnaround in global emissions, the tub will be filling up further for quite a while.

There's always the hope that some technological fix will sweep in and magically close the faucet or open up the drain. That's unlikely, to say the least. The technological fix most often bandied about—geoengineering—is sometimes seen to have a flavor of that: turn down the sun a bit to cool the planet. It's really anything but. It treats the symptom—the resulting higher temperatures—without tackling the root cause. It would affect neither the faucet nor the drain and would do nothing about actual water levels. And trying to manage one pollution problem by shooting more pollution into the stratosphere could have vast unintended consequences.

One technological fix that may well help open up the drain in a big way is directly removing carbon from the atmosphere. It comes under various guises: "air capture," "direct carbon removal" (DCR), or "carbon dioxide removal" (CDR). Confusingly, some also call it "geoengineering." That's really a misnomer. It has all the characteristics of addressing the root cause of the problem, rather than a techno fix to treat the symptoms. That's good, though it also means it's slow and—at least so far—prohibitively expensive. Some companies are beginning to file patents to take advantage of a world that prices carbon and creates incentives for this technology. But price carbon we must. Otherwise it won't happen. Burning carbon + removing carbon is, by definition, more expensive than just burning carbon.

BREAKTHROUGHS

It's why humans left their caves, domesticated animals, invented the wheel, built cities, got behind the wheel—and ended up sitting in a chair in the air, flying across oceans, streaming movies on their iGadgets. We live longer than ever, more comfortably than ever, all thanks to human ingenuity at work. Breakthroughs will save the day again.

Maybe. Or maybe not. The pace of innovation today is so fast and so unprecedented in human history that it's tough to take the past as guidance. And there, too, are signs pointing in both directions.

Some pollution problems go away because the one company holding the patents for the offending pollutants invents an environmentally friendly replacement, and even then it takes concerted government action (see "Protocol, Montreal"). Some pollution problems don't ever seem to go away (see "Protocol, Kyoto"). But sitting around waiting for a breakthrough is hoping for the best. We need to prepare for the worst. In fact, we can do a lot better than praying for salvation. Both the problem and the solution are staring us in the face.

CARBON DIOXIDE

The problem. The main one, at least.

Other greenhouse gases like methane and powerful industrial gases like hydrofluorocarbons (HFCs) as well as black carbon have significant influence under much shorter time scales—years or maybe a decade or two. The *level* of eventual warming, however, is most closely linked to carbon dioxide.

Technically, water vapor has an even larger, total effect. But that's beside the point. It's not something humans control directly. The chemical chain goes from carbon dioxide to higher temperatures to more water vapor. In the end, it still comes down to carbon dioxide.

CARBON PRICE

The solution. The main one, at least.

There are many others, though most attempt to mimic a price on greenhouse gas pollution in one form or another.

Some manage to do that relatively directly and cheaply. Others are more expensive, though also more opaque and therefore sometimes easier to swallow politically. None, however, is typically quite as efficient as rechanneling fundamental economic forces directly—capping or taxing carbon in the first place.

And there's another important opportunity here that often gets lost in the shuffle: subsidizing low-carbon technologies.

CHANGE, DIRECTED TECHNICAL

Putting the first solar panel on a roof takes time and money. The millionth is quick and cheap. The trick is to get over the initial hump. The best policy: subsidize innovation—or more specifically, "learning-by-doing."

California's Solar Initiative is a good example of just such a policy: it subsidizes the installation of panels early on and pulls support back almost immediately. Independent analysis suggests that it has hit the mark.

Not all subsidies are good. They often tend to be abused and misused. Once introduced, they tend to stick around long past their useful lifespan. The $500 billion global fossil fuel subsidies are a good example. There may well have been a good reason for them once—long before the world woke up to the climate problem. The same reasons that now apply to clean technologies once applied to oil, coal, and gas: large potential benefits facing seemingly insurmountable hurdles. Entrenched interests, say the horse-and-buggy or whale oil lobbies, do everything to undermine a new industry getting a foothold. Sooner or later, though, things turn. Market and government interests begin to favor the new industry. That's around the time when subsidies ought to end. "Infant industry protection" no longer applies. If anything, the task now shifts to breaking up outsized monopolies, as happened a century ago with Standard Oil.

These caveats notwithstanding, positive learning-by-doing spillovers are as real as the negative carbon spillovers. The solution to correct the carbon spillover is clear. (See "Carbon Price.") The solution to correct the learning-by-doing spillover is subsidies. The overall process comes under the heading: directed technical change.

CLIMATE SCIENCE

Year by which the greenhouse effect had been discovered: 1824.

Year by which the greenhouse effect had been shown in a lab: 1859.

Year by which the greenhouse effect had been quantified: 1896.

Year by which today's range for the all-important "climate sensitivity" metric had been established: 1979.

CLIMATE SENSITIVITY

Double carbon dioxide concentrations in the atmosphere, and global average temperatures are bound to increase as well. The level of increase is called "climate sensitivity." Trouble is we don't know the exact number.

Despite plenty of advances in climate science, the range for climate sensitivity has been 1.5 to 4.5°C (2.7 to 8°F) seemingly forever, at least since 1979. Confidence in the range itself has increased, and a brief interlude from 2007 to

2013 even narrowed it to 2 to 4.5°C (3.6 to 8°F). Low temperatures seemed to be out for a while. Bad news all around. Adding them back in, however, isn't exactly good news. It just means that the deep-seated uncertainties are even more profound than previously thought. The exact value of climate sensitivity may never be resolved completely. Or rather, once we know its precise number—hundreds of years into the future—it will be too late to do us any good.

To add to the sense of insecurity, the 1.5 to 4.5°C (2.7 to 8°F) range is only the "likely" temperature effect of doubling carbon dioxide concentrations. We can expect the final number to fall somewhere in that range, but that's far from certain. The overly precise definition of what "likely" means says there's at least a 66 percent chance of hitting that range. The flipside is that it implies an up to 34 percent chance of it being either below or above, with a lot more room above. That's where deep uncertainty starts to bite. It's also where we'll ask you to wait. The next chapter, on "Fat Tails," is all about it.

DICE

Given the enormous uncertainties inherent in the climate prediction business, it might be easy to conclude that we just can't tell. That's clearly not good enough. Bill Nordhaus's Dynamic Integrated Climate-Economy (DICE) model is the most prominent of those that try to make sense of it all. It takes trade-offs between the climate and the economy as a starting point to calculate an optimal path and price for carbon dioxide emissions.

In many ways, DICE & Co are simply tools. It's up to others to make the assumptions that spit out an optimal carbon price. Nordhaus's own preferred assumptions come up with a price of around $20 per ton of carbon dioxide emitted today.

Perhaps the best current number comes courtesy of a major coordinating effort by the U.S. government. That price is around $40 per ton emitted today, derived from averaging three models, including DICE. That's a good start, but it's still far from assessing the full costs of global warming.

The underlying models do their best to capture the "known knowns," and even there they miss quite a bit. By definition, they don't yet capture the "known unknowns." And as so often is the case, it may well be the "unknown unknowns" that define the final outcome. The $40 then can only be considered a lower bound for the social cost of carbon. Most of what's left out will increase the number further still.

EXTERNALITIES

An economist's way of saying "problem." It's when markets—left to their own devices—fail. Externalities come in two flavors: positive and negative.

Learning-by-doing is a good example of a positive externality. Without added incentives, inventors don't consider the fact that their inventions contribute to the greater good, and, thus, they invent too little. (See "Change, Directed Technological.")

Climate change is the mother of all negative externalities. Seven billion of us emit tens of billions of tons of carbon dioxide into the atmosphere every year. The costs are large—at least $40 per ton emitted—but those doing the polluting aren't paying for it directly. (See every other part of this book, not counting the bit on "Free Drivers".)

FREE DRIVERS

Carbon dioxide is the problem. Pricing it properly is the solution. Then there's: "The deliberate large-scale manipulation of an environmental process that affects the earth's climate, in an attempt to counteract the effects of global warming." That's the *Oxford English Dictionary*'s definition of geoengineering, and it's as good as any. Some include various attempts at taking carbon dioxide out of the atmosphere. We don't.

Instead, when you hear "geoengineering," think about something closer to a volcanic eruption: shooting sulfur dioxide (and, in the case of volcanoes, lots of other gunk) into the stratosphere to reflect sunlight and lower temperatures. The 1815 eruption of Mount Tambora brought us the "year without a summer." By some accounts, it led to 200,000 deaths across Europe in 1816. By other accounts, it forced Mary Shelley and John William Polidori to spend much of their Swiss summer holiday indoors, leading to both *Frankenstein* and *The Vampyre*. By still other accounts, the latter has since morphed into *Dracula*.

Actual geoengineering proposals have little to do with violent volcanic eruptions, or with *Frankenstein* and *Dracula*. Most geoengineering proposals concern controlled, small-scale attempts at counteracting the global temperature increases by spewing sulfuric acid vapor or other tiny sulfur particles into high altitudes. This type of geoengineering has one crucial characteristic: It's cheap—at least in the narrow sense of costs incurred by those doing the "shooting."

Geoengineering has many potential problems, but cost isn't one of them. Welcome to the father of all negative externalities, a.k.a. the *free-driver* effect.

It would be so cheap to crudely geoengineer the planet's temperature that one person, or more likely one country's concerted research effort, could do it. It wouldn't take much to commandeer a small fleet of high-flying planes capable of delivering sulfur into high altitudes year after year. Volcanoes do this naturally. Mount Pinatubo, which last erupted in 1991, cooled global temperatures by about 0.5°C (0.9°F) the following year. Unless the world acts to control global warming pollution in the first place—and possibly even if it does—we may well be looking at a geoengineered planet sometime soon.

This combination of low direct cost and high leverage makes the *free-driver* effect almost the exact opposite of what is causing the problem in the first place.

FREE RIDERS

The heart of the global problem that is global warming. It's beggar-thy-neighbor to the extreme, except that all seven billion of us are neighbors. Why act, if your actions cost you more than they benefit you personally? Total benefits of your actions may outweigh costs. Yet the benefits get spread across seven billion others, while you incur the full costs. The same logic holds for everybody else. Too few are going to do what is in the common interest. Everyone else free-rides.

Smaller free-rider problems can indeed be solved through community engagement and other informal arrangements. Elinor Ostrom won a Nobel Prize for that insight. Alpine Swiss farmers have been bringing their cows to common pastures for centuries, without overexploiting their shared resource. The secret is that the farmers don't just compete for poorly defined grazing rights; they also meet at the market, at school, and at church. They know the people their actions affect. Something similar goes for village aquifers, community-managed fisheries, and plenty of other examples of potential free-rider problems at relatively small and ultimately manageable scales.

Global warming is different. Hundreds of thousands—even millions—of committed environmentalists striving to minimize their own carbon footprints can't do much by themselves. Those committed enough to the cause can try to decrease their personal emissions to zero—something that's neither possible nor, ultimately, desirable, given current technologies—and it will still be far from what we need. The numbers don't add up. They'll only begin to add up when environmentalists use their collective political powers to move the policy needle in the right direction, toward a price on carbon.

IRREVERSIBLE

A relative concept. Very little truly is. But climate change operates on such large timescales—decades and centuries—that many of the effects might as well be irreversible. Elevated levels of carbon dioxide remain in the atmosphere for centuries and millennia. Getting them down is extremely difficult. Think about the bathtub: enormous amounts of carbon dioxide in the atmosphere; small drain.

All that magnifies other irreversibilities like sea-level rise. Yes, the poles have been ice-free before. And even if they turn ice-free this time around, they would surely freeze again, if carbon dioxide concentrations were lowered eventually to preindustrial levels. But that freezing wouldn't happen for centuries or millennia, too far into the future to matter to those affected now or in the next few generations. Greenland and the West Antarctic ice sheet alone hold enough water to raise global sea levels by over 10 meters (30 feet). That directly affects at least 10 percent of the global population, and indirectly it affects pretty much everyone else.

LAW OF DEMAND

Price goes up, quantity demanded goes down. That's as close to a "law" as economists will ever get.

There's only one seeming exception to that rule, and it's only ever been found to apply in situations when you are so poor that the fear of starvation determines what you eat. Poor, rural Chinese in the South of China eat lots of rice. Like almost everyone else, they'd like to eat more meat as they become richer. But when the price of rice goes up, some eat less meat and more rice with veggies instead. Rice in this case then is what's become known as a "Giffen good," after Sir Robert Giffen, who reportedly observed something similar among the Victorian era poor. For our purposes, none of this has much relevance.

The quantity demanded for almost all goods does indeed go down when price goes up. That relationship is also true for some clear *bads*—it's why we tax cigarettes. And it points to the most obvious of solutions: price carbon. As the price of emitting carbon rises, carbon emissions fall.

OCEAN ACIDIFICATION

An oft-ignored yet crucial consequence of rising carbon dioxide levels in the atmosphere. Most of the carbon dioxide emitted eventually ends up in the oceans, turning them more acidic.

Oceans are already over 10 percent more acidic than they were around 1990, when measurements began in earnest. They're likely over 25 percent more acidic than at the dawn of the industrial revolution. That's less of a total increase—percentage-wise—than the 40 percent increase in carbon dioxide levels in the atmosphere, but small changes in acidity can make a big difference. And changes in acidity are no longer small. Ocean acidity now is increasing ten times faster than during the last great die-off of some marine organisms around 56 million years ago—when the world moved from the Paleocene to the Eocene. The proximate cause back then? A sharp increase in carbon dioxide, and sudden global warming of around 6°C (11°F).

Too little is known about the full implications of more acidic oceans. The partial marine die-off 56 million years ago isn't commonly seen to be on the same level as the giant asteroid that wiped out the dinosaurs 65 million years ago. Some marine organisms were killed; others thrived. That's scant assurance. More acidic waters, for example, are particularly bad news if you are a shellfish because they prevent you from forming your shell in the first place. It's tough though to pinpoint the larger, possibly even more consequential effects.

Enter "alkalinity addition," one type of geoengineering method proposed as a direct response. Adding enough calcium carbonate powder—a.k.a. ground-up limestone—to the oceans would bring down their acidity. The trouble is that taking this step further increases ocean uptake of carbon dioxide, and the cycle continues. An even bigger trouble: any of the methods currently discussed tend to be rather expensive, very much unlike the free-driver effect inherent in Mount Pinatubo–style geoengineering to lower global temperatures. Instead, they are much closer to what it would cost to reduce carbon dioxide emissions in the first place, if not higher. Ipso facto, why not focus on reducing carbon dioxide instead? Doing so tackles both the ocean acidification and the temperature problem.

PROTOCOL, KYOTO

Technically, the "Kyoto Protocol to the United Nations Framework Convention on Climate Change."

Few have heard the full name, though it is significant. The latter Framework Convention part is really where the legal action is. UNFCCC—pronounced "U, N, F, triple-C"—came out of the 1992 Earth Summit in Rio de Janeiro. It was ratified by 195 nations, including the United States. It gave us the infamous phrase: "stabilize greenhouse gas concentrations in the atmosphere at a level that would prevent dangerous anthropogenic interference with the climate system." Despite the fact that the Framework Convetion is a binding treaty, that goal seems to be long gone, and it's a bit unclear whom you could sue if, say, your island nation is disappearing under the rising seas.

The Kyoto Protocol itself, as opposed to the UNFCCC, is a different matter. For one, the United States signed but never ratified it. Canada did but has since pulled out. The European Union did and is sticking to its ambitious climate goals. Sadly, that's far from enough to solve the global problem, especially considering the fact that China and India have no formal emissions reductions commitments under Kyoto.

A global problem, in the purest form, requires a global solution. That means having sensible climate policies in place to guide seven billion people onto a more sustainable path. The glimmer of hope, as it were, is that it doesn't take nearly 195 countries all committing to their own, strong policies to make a dent. There are a gazillion ways to slice and dice the problem, but it's really just a handful of major polluters that account for the majority of global greenhouse gas emissions. Add up the United States, Europe, China, India, Japan, Russia, and a bit of Brazil and Indonesia—the large forestry bit, that is—and you get to over 60 percent of global emissions. Now, how to solve each of these pieces is a different matter, but at least it's not a 195-piece, multidimensional jigsaw. In fact, there are hopeful signs in each of these countries and regions that serious policy change is not a matter of "if" but "when." Time, of course, is the all-important factor, and the task at hand is to hasten the march of history.

PROTOCOL, MONTREAL

Technically, the "Montreal Protocol on Substances that Deplete the Ozone Layer."

This protocol, signed in 1987, is widely considered to be one of the biggest environmental success stories. Books have been written about the reasons for its success. The full story is complex, but a simple telling goes like this: DuPont, which held many of the patents on the gases that were destroying the ozone layer, discovered commercial opportunities in using alternatives that were safer for stratospheric ozone. The profit motive found a happy union with feel-good public relations. Once DuPont decided to change its position by 180 degrees virtually overnight, it didn't take long for the U.S. government under none other than President Ronald Reagan to change its tune and pass and ratify the international treaty that helped phase out the offending gases. The ozone hole has now been receding for years and is expected to heal completely by midcentury. Crisis averted.

That's good for the ozone layer. By now we know it's also bad for the climate. The Montreal Protocol regulates chlorofluorocarbons (CFCs) and hydrochlorofluorocarbons (HCFCs). DuPont discovered that they are readily replaced by hydrofluorocarbons (HFCs). Unfortunately, HFCs are a hundred to over 10,000 times as potent as carbon dioxide when it comes to global warming. Good thing we only use it in small quantities, but that doesn't mean its use shouldn't be reduced. Quickly. Unfortunately, the control of HFCs doesn't fall under the Montreal Protocol (yet) but instead under Kyoto. Back to square one.

The success of the Montreal Protocol does point to one important lesson: change is, in fact, possible. It might be more difficult with climate than with the ozone hole, but that doesn't mean climate change is an uncontrollable, runaway problem. It requires concerted effort and unprecedented political leadership, but there's a good reason why every prediction of where the climate is heading necessitates a disclaimer: here's what will happen, *unless society changes course*. The ozone hole didn't turn out to be as bad as some of the best scientists had predicted. That's not because those scientists were wrong, but because the world actually came together to do something about the problem before it got too severe.

The same goes for climate change. The world could take control, if it only wanted to. Policy could—*should*—guide humans to render the most devastating climate predictions wrong, precisely because society will have reacted to the most dire of warnings.

TRADE-OFFS

A concept that's ingrained in any economist's DNA. Hardly ever are there true absolutes. Outright bans can sometimes make sense, as the example of the Montreal Protocol effectively banning CFCs shows. But bans often entail high costs. Banning all carbon dioxide emissions is out. It would simply be too costly.

One of the most relevant trade-offs to us here is most simply summarized as "growth versus climate." After all, there's no denying that economic growth of the kind Europe and the United States have experienced since the industrial revolution and that China, India, and many others are experiencing now comes with unaccounted-for costs. Perhaps the largest such cost is unabated climate change.

The flipside of this growth-climate trade-off is that there must be an optimal path that puts the benefits and costs of each into balance. In theory, there is. The big question is whether, in practice, there is indeed a way to make sense of it all in one, comprehensive benefit-cost analysis. See "DICE" for a model that attempts to do just that. But what if uncertainties are so large as to make a mockery of any such dollar estimate?

UNCERTAINTY

Think back to the "likely" range for climate sensitivity, where global average temperatures will likely end up once concentrations of carbon dioxide double. It may well be the "unlikely" numbers outside that range that define the final outcome. And the news there isn't good.

CHAPTER 3

Fat Tails

I N 1995, the Intergovernmental Panel on Climate Change (IPCC) declared it was "more likely than not" the case that global warming was caused by human activity. By 2001, it was "likely." By 2007, it was "very likely." By 2013, it was "extremely likely." There's only one step left in official IPCC lingo: "virtually certain." The big question is how certain the world needs to be to act in a way that is commensurate with the magnitude of the challenge.

An equally important question is whether all this talk of certainty is conveying what it ought to convey. The increasing likelihood of anthropogenic climate change has three sides to it. Only one of them is good.

The first piece of bad news is that we humans are, in fact, increasing global temperatures and sea levels alike. It would have been cause for celebration if, say, the 2013 report decided that science had gotten it wrong all along. Imagine the *New York Times* headline: "IPCC says decade 'without warming' here to stay." Alas, no such luck. Modern atmospheric science once again confirmed the basic ideas of high school chemistry and physics, going back to the 1800s: more carbon dioxide in the atmosphere traps more heat.

The good news then, in some twisted philosophical sense, is the confirmation of bad news. Climate science has progressed over the past couple decades to the point where it is now able to make the definitive statement that global warming is *extremely likely* caused by human activity. We know enough to act. Ignoring that reality, by now, would amount to willful blindness.

But there's an additional piece of bad news: the false sense of security conveyed by all this talk about certainty. At least by one important measure, we don't appear to be closer to understanding how much our actions will warm the planet than we were in the 1970s, at the dawn of modern climate science and long before the first IPCC report. Worse still is that what we have learned since then points to the fact that what happens at the very extremes—the tails of the distribution—may dwarf all else.

SENSITIVE CLIMATES

In 1896—eight decades before Wally Broecker coined the term "global warming," and long before anyone knew what a climate model was—Swedish scientist Svante Arrhenius calculated the effect of doubling carbon dioxide levels in the atmosphere on temperatures. Arrhenius came up with a range of 5 to 6°C (9–11°F). That effect—what happens to global average surface temperatures as carbon dioxide in the atmosphere doubles—has since become known as "climate sensitivity" and has turned into an iconic yardstick.

Climate sensitivity itself is already a compromise, a way of making an incredibly complex topic slightly more tractable. The parameter does have a few things going for it. For one, the starting level of carbon in the atmosphere doesn't matter, at least not by much. One of the few rather well-established facts is that eventual global average temperatures scale linearly with percentage changes in underlying carbon dioxide concentrations. The first 1 percent increase in carbon in the atmosphere has a similar impact as the 100th. Any doubling of concentrations, from anywhere within a reasonable range, leads to roughly the same eventual increase in global temperatures. The definition of climate sensitivity plays off that fact.

A doubling of preindustrial carbon dioxide levels of 280 parts per million (ppm) seems all but inevitable. The world has just passed carbon dioxide concentrations of 400 ppm, and levels are still rising at 2 ppm per year. Counting other greenhouse gases, the International Energy Agency (IEA) estimates that the world will end up somewhere around 700 ppm by 2100—two-and-a-half times preindustrial levels—unless major emitters take drastic additional steps.

Luckily, Arrhenius's climate sensitivity range of 5–6°C (9–11°F) has proven to be too pessimistic. In 1979, a National Academy of Sciences Ad Hoc Study Group on Carbon Dioxide and Climate concluded that the best estimate of climate sensitivity was 3°C (5.4°F), give or take 1.5°C (2.7°F).

"Conclude" may be a bit strong a term to use here. The process is commonly retold thus, not without admiration for academic genius at work: Jule Charney, the study's lead author, looked at two prominent estimates at the time—2°C (3.6°F) on one end and 4°C (7.2°F) on the other—averaged them to get 3°C (5.4°F), and added half a degree centigrade on either end to round out the range because, well, uncertainty.

Thirty-five years of ever more sophisticated global climate modeling later, our confidence in the range has increased, but what's now called the "likely" range of 1.5 to 4.5 °C (2.7 to 8°F) still stands. That should be a tipoff right there that something rather strange is happening. There's something stranger still.

A PLANETARY CRAPSHOOT

The IPCC defines "likely" events to have at least a 66 percent chance of occurring. That still tells us nothing about whether things may turn out all right—with climate sensitivity closer to 1.5°C (2.7°F)—or not all right at all—closer to 4.5°C (8°F). Taking the IPCC probability descriptions literally, the chance of being outside that range would be up to 34 percent. There's no precise verdict as to where these 34 percent go, though there's clearly more room above 4.5°C (8°F) than below 1.5°C (2.7°F). See figure 3.1.

For any numbers below 1.5°C (2.7°F), we could rightfully celebrate—ideally with a bottle of Champagne, flown in from France for the occasion, and letting out an extra puff of carbon dioxide when opened. Though that's unlikely. And not even these low climate sensitivity realizations of 1.5°C (2.7°F) provide a guarantee that climate change won't be bad. In fact, quite the opposite: At 700 ppm, final temperatures would still rise to higher than where they were over three million years ago. Think

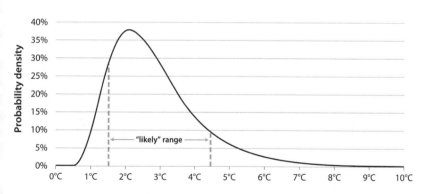

Figure 3.1 **Eventual global average surface warming due to a doubling of carbon dioxide (climate sensitivity)**

back to camels in Canada, who happily marched in what is now frozen tundra at temperatures of 2–3.5°C (3.6–6.3°F) above preindustrial levels. And we'd be at 2°C (3.6°F) with a climate sensitivity of 1.5°C (2.7°F), which is at the lower edge of the likely range.

All that makes our inability to exclude climate sensitivities above 4.5°C (8°F) all the more significant. Any probability of climate sensitivity that high should make for (heat-induced) shudders. The most important question then is: how fast does the chance of hitting any of these higher climate sensitivity figures go to zero as the upper bound of climate sensitivity increases? One could imagine an extreme scenario in which the chance that climate sensitivity is above 4.5°C is greater than 10 percent, but if the chance of being above 4.6°C were zero, we could exclude any even higher numbers. If only the planet were that lucky. It's *extremely unlikely*—in the English rather than the strict IPCC sense of that term—that the probabilities of higher climate sensitivities would drop off that quickly.

It's much more likely that the chance of hitting higher temperatures tapers off at an uncomfortably slow pace, before hitting something close enough to zero to provide a reasonable level of comfort that even more extreme numbers won't materialize. That scenario is closer to what statisticians describe as a "fat tail." The probability of 4.6°C is smaller than for 4.5°C, though not by much.

The all-important question, then: how likely is a potentially catastrophic realization of climate sensitivity? The IPCC says it's "very unlikely" that climate sensitivity is above 6°C (11°F). That's comforting but for its definition of just what "very unlikely" means: a chance of anywhere between 0 and 10 percent. And that range is still only the likelihood that *climate sensitivity* is above 6°C (11°F), not actual temperature rise.

||||||||||||||||||

Let's jump right to the conclusion. Take the latest consensus verdict at face value and assume a "likely" range for climate sensitivity of between 1.5 and 4.5°C (2.7 and 8°F). Equally important, stick to the IPCC definition of "likely" and assume it means a chance of greater than 66 percent, but less than 90 percent. (The latter would be "very likely.") And take the IEA's interpretation of current government policy commitments at face value. Here's what you get: about a 10 percent chance of eventual temperatures exceeding 6°C (11°F), unless the world acts much more decisively than it has.

Figure 3.2 and table 3.1 are the culmination of parsing umpteen scientific papers and countless hours spent fretting over how to get it just so. Rows 1 and 2 in the table

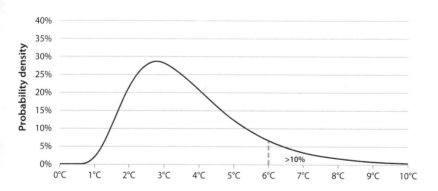

Figure 3.2 **Eventual global average surface warming based on passing 700 ppm CO_2e**

Table 3.1. Chance of eventual warming of >6°C (11°F) rises rapidly with increasing CO_2e concentrations

CO_2e concentration (ppm)	400	450	500	550	600	650	700	750	800
Median temperature increase	1.3°C (2.3°F)	1.8°C (3.2°F)	2.2°C (4.0°F)	2.5°C (4.5°F)	2.7°C (4.9°F)	3.2°C (5.8°F)	**3.4°C** **(6.1°F)**	3.7°C (6.7°F)	3.9°C (7.0°F)
Chance of >6°C (11°F)	0.04%	0.3%	1.2%	3%	5%	8%	**11%**	14%	17%

represent the move from carbon dioxide–equivalent (CO_2e) concentrations in the atmosphere to ultimate temperature increases. Row 3 shows the corresponding chance of exceeding final average temperature increases of 6°C (11°F). Whenever we had to make a judgment call of where to go next, we tried to take the more conservative turn, which may well underplay some of the true uncertainties involved.

The scariest bit is just how fast the chance of eventual temperatures exceeding 6°C (11°F) goes up. Compare changes in the median temperature increase with the chance of passing 6°C (11°F). Going from 400 to 450 ppm, the difference between 1.3°C (2.3°F) and 1.8°C (3.2°F) for the most likely temperature increase, may not be all that much. There may be some, potentially irreversible, tipping points along the way, but ultimately it's only half a degree centigrade (less than a degree Fahrenheit), or an increase of a bit more than a third. At the same time, the chance of exceeding 6°C (11°F), the last row, just jumped from 0.04 percent to 0.3 percent, almost tenfold. All that's just for moving from 400 to 450 ppm, while the world has already passed 400 ppm for carbon dioxide alone and 440 to 480 ppm for carbon dioxide–equivalent concentrations! A further jump to 500 ppm increases that chance of catastrophe to 1.2 percent. By the time concentrations reach 700 ppm— where the IEA projects the world will end up by 2100 even if all governments keep all their current promises—the chance of eventually exceeding 6°C (11°F) rises to about 10 percent. That looks like the manifestation of a fat tail, if there ever was one (even though strictly speaking we don't even assume that property in our calculations; our tail is "heavy," not quite "fat" in statistical terms).

At 700 ppm, the median temperature increase would be 3.4°C (6.1°F). This alone would be a profound, earth-as-we-know-it-altering change. Polar regions would likely warm by at least twice that global average, with everything that entails. The costs would be staggering and should have prompted the world's leaders to head off such a possibility long ago. Yet those costs are still nothing compared to what would happen if final temperatures were to exceed 6°C (11°F). It's the roughly 10 percent chance of near-certain disaster that makes climate change costlier still.

Now we are truly in the realm of what Nassim Nicholas Taleb describes as a "Black Swan" and Donald Rumsfeld as "unknown unknowns." We don't know the full implications of an eventual 6°C (11°F) temperature change. We can't know. It's a blind planetary gamble. Devastating home fires, car crashes, and other personal catastrophes are almost always much less likely than 10 percent. And still, people take out insurance to cover against these remote possibilities, or are even required to do so by laws that hope to avoid pushing these costs onto society. Risks like this on a planetary scale should not—*must not*—be pushed onto society.

"Must not" is a strong phrase. It conjures images of bans or—in dollars and cents—infinite costs. That goes head-to-head against any economist's idea of trade-offs. The costs of global warming may be high, perhaps higher than anyone thought possible. But surely, they can't be infinite.

MONEY IS EVERYTHING

Trying to estimate the eventual temperature increase is one thing. But even if we knew that number with any precision for that one hot day in August 2100 in Phoenix, what actually concerns us isn't necessarily how high temperatures climb. We care more about climate impacts, and how much they will cost society. Sea-level rise is one. Another is extreme events like droughts or hurricanes that might hit your home long before rising sea levels would drive you from it altogether.

The business of pinning down specific impacts is messy and fraught with its own uncertainties. There are known unknowns aplenty. Unknown unknowns may yet dominate. And tipping points and other nasty surprises seem to lurk around every corner. Some of them may put warming itself on overdrive. Releasing vast carbon deposits in Siberian or Canadian permafrost could prove to be a tipping point resulting in bad global warming feedbacks. Others may have relatively less influence on actual temperatures but have plenty of other impacts. Melting of Greenland and the West Antarctic ice sheets alone already raises sea levels by up to one centimeter (0.4 inches) each decade. If the Greenland ice sheet fully melted, sea levels would rise 7 meters (23 feet). Full melting of the West Antarctic ice sheet would add another 3.3 meters (11 feet). That's not happening tomorrow or even this century. The IPCC's estimates of global average sea-level rise for this century top out at 1 meter (3 feet). But the tipping point at which the full eventual melting becomes inevitable will be passed much sooner. We may have already passed the tipping point for the West Antarctic ice sheet.

These compounding uncertainties—first from emissions to concentrations to temperatures, and then from temperatures to ultimate impacts measured in dollars and cents—make things extremely hard to get right. That hasn't stopped economists from trying.

One of the best is Bill Nordhaus. His DICE model—short for Dynamic Integrated Climate-Economy model—has been publicly available since the early 1990s. Generations of graduate students have played around with it, tried to poke holes in it, and derived estimates of "optimal" global climate policy. Nordhaus's own estimates of the social cost of carbon have been going up ever since the model was first released in 1992. Back then, his economically optimal response to climate change was a global carbon tax of about $2 per ton of carbon dioxide (in 2014 dollars). That went hand in hand with global average warming climbing to 4°C (7.2°F) and beyond. In the tug-of-war between economic growth and a stable climate, growth won. Climate impacts have been catching up ever since, pushing unfettered, fossil-fueled growth further and further from being optimal. Today, Nordhaus's preferred "optimal" estimate is around $20 per ton of carbon dioxide. The resulting final temperature increases now top out at around 3°C (5.4°F).

The search for the optimal carbon price is a hot button issue. Nordhaus's formally derived $20 is lower still than the average estimate of $25 per ton, presented in his own book as an "illustrative" example. That, in turn, is lower than the current "central" U.S. government's estimate of around $40, derived from a combination of outputs from DICE and two other assessment models.

None of that yet factors in the proper cost of the tails, fat or otherwise. Nordhaus's maximum average temperatures may stay below 3°C (5.4°F), but that's the average. It still leaves unspecified the probability of topping 6°C (11°F) or more. Some other estimates attempt to take uncertainty more seriously. The U.S. government itself presents what it calls the "95th percentile estimate" as a proxy of sorts for capturing extreme outcomes. The optimal number there: over $100 per ton of carbon dioxide emitted today.

What then does the central $40 estimate include, and how is it derived? Two key issues loom large: dollar estimates of damages caused, and discounting. We'll address them in turn.

HOW MUCH FOR A DEGREE OF WARMING?

Compare the average climates in Stockholm, Singapore, and San Francisco. Winters in Sweden are long, cold, and dark. You'll have to wait for the summer months to get average highs above 20°C (68°F). Singaporeans don't have this problem. Their average low is higher than Stockholm's average high year-round. All that makes San Franciscans feel smug, fog and all. They enjoy stable Mediterranean climates year-round, with a week of rain in "winter." Still, all three cities are thriving metropolises. Historians may even argue that all of them got their start because of winning geographies. What, then, should lead us to believe that one climate is better or worse than another? Or that warmer average global temperatures come with costs?

The costs of climate change aren't the result of moving away from some mythical optimal climate. Stockholm may be a more pleasant place with a degree or two extra. Incidentally, that's precisely what Swedish scientist Svante Arrhenius, of greenhouse effect fame, suggested we may want to do deliberately: burn more coal "to enjoy ages with more equable and better climates, especially with regards to the colder regions of the earth." In Arrhenius's defense, he said so in 1908, after he had identified the greenhouse effect, but long before it became clear that there are significant costs to pumping carbon dioxide into the atmosphere. In the end, the costs of small temperature changes are, for the most part, the sum of the costs of changing what we've gotten used to. And it's not just that Swedes already own winter jackets and Singaporeans air conditioners. It's massive investments and industrial infrastructures, built around current climates—and current sea levels—that make temperature increases costly.

And once again, it isn't temperatures themselves that matter as much as what these rising temperatures entail. One such effect is rising sea levels. Then there are storm surges on top—by then stronger and more frequent precisely because of climate change. And all that's the perfectly "normal," *average* effect of sea-level rise baked into where we *are* already heading. None of this is yet taking into account fat tails or other catastrophic scenarios.

When models incorporate the latest science and quantify ever more of the damages likely to occur because of climate change, the estimated costs of carbon pollution go up. DICE & Co are perennially playing catch-up with the latest science. In 2010, the central U.S. government's estimate of the social costs of one ton of carbon dioxide emitted in 2015 was around $25. The 2013 iteration increased it to around $40.

|||||||||||||||||

None of this is meant to decry the modeling efforts. Quite the opposite. Getting things right is incredibly difficult. If anything, it is a call to invest in economic modeling—in a big way. Nordhaus's DICE model, as well as its main competitors, FUND and PAGE, were all started by one person, and have been painstakingly maintained, patched, and modified over years and decades by a small group of dedicated economists. Meanwhile, when big business tries to analyze what toothpaste flavor to sell where, it uses massive quantities of geo-spatial, customer-level data, analyzed by dozens of dedicated statisticians and programmers.

We certainly shouldn't scrap economic climate models for their inadequacy. If anything we should be supercharging them: IBM-ifying their operation. There's much more at stake here than with selling toothpaste. Yet Colgate and Procter & Gamble are competing with the help of massive data operations, while DICE can run on your home computer. More manpower and data would at least help the models incorporate the latest available information in real time.

Even if we did all of that, though, there would still be one major problem: How should we quantify the damages caused by potentially catastrophic climate change? More data won't necessarily help us make inroads on that question.

DICE & Co mostly look to the past for guidance. Hundreds of scientific studies try to quantify the impacts of global warming on anything from sea-level rise to crop yields to tropical storms to war. The task then is to translate those impacts into dollars and cents. We quickly run into two problems. For one, only a small part of known damages can be quantified. Lots are missing. The list of currently unquantified and—at least in part—unquantifiable damages spans everything from known respiratory illnesses from increased ozone pollution due to warmer surface temperatures to the effects of ocean acidification. Moreover, the only parts we can truly quantify are in a relatively narrow, low-temperature range: changes of fractions of a degree, maybe 1°C (1.8°F), or maybe 2°C (3.6°F) of global average warming. How can we estimate what happens at 5, 4, or even 3°C (9, 7.2, or 5.4°F)?

Extrapolate, extrapolate, extrapolate.

That's at least what current models do. Take what happens at 1 or 2°C and scale it up. We know that, because of tipping points and other possibly nasty surprises, we can't just look at things linearly. No one seriously proposes that. Instead, DICE mostly relies on something close to quadratic extrapolations: If 1°C causes $10 worth of damages, then 2°C doesn't cause $20—that's linear—but $40. More specifically, Nordhaus estimates that warming of 1°C costs less than 0.5 percent of global GDP, 2°C costs around 1 percent, 4°C costs around 4 percent. Things take off after that, but even 6°C stays below 10 percent.

Mind you, that's a big absolute number: 10 percent of total, global economic output today would be around $7 trillion. If they were to materialize, by the time these 6°C (11°F) changes were to hit a century or more from now, the fraction of damages would be multiplied by a large growth factor. But how can we be certain that it's the right number? We can't. Once we extrapolate damage estimates

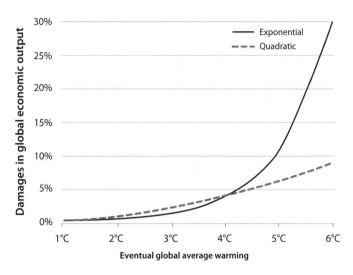

Figure 3.3 **Quadratic and exponential extrapolations of global economic damages**

as far out as 6°C (11°F), it all becomes guesswork. Using a quadratic function is a convenient shortcut, but it's not much more than that. Lots of other extrapolations would fit the observed damages on the lower end of the scale but would yield wildly different results on the upper end. For instance, figure 3.3 shows how estimating exponential rather than quadratic warming yields starkly different results:

For 1°C and 4°C, the two lines are identical. For 2°C and 3°C, they are close enough to be indistinguishable given the uncertainties. At 5°C, things begin to diverge. By 6°C, they might as well be describing different planets. The quadratic extrapolation ends up at a bit under 10 percent of global economic output. The exponential comes in closer to 30 percent.

We aren't saying a 30 percent decline in output is any more correct than 10 percent decline should global average temperature increases hit 6°C (11°F). We just don't know. And no one else does either. One could tell stories about how 10 percent may be too high because people will be able to cope. Even with 6°C (11°F) of warming, Stockholm then will still be cooler than Singapore now. Or one could tell stories around how 30 percent may still be too low because neither Stockholm nor Singapore would be around to see the day. Their current coast lines would be on track to be submerged under several meters of water. *Would*, not *may*. But it's once again the deep-seated uncertainties around the eventual extent and timing of the consequences that add to the true costs.

||||||||||||||||||||

It's also not at all clear that we should be thinking about damages as a percentage of output in any given year. Standard practice for DICE and other models is to assume that the economy hums along just fine until damages from climate change get subtracted at some point in the future. Catastrophic or not, conventional estimates of climate damages will feel small compared to the amazing increases in wealth that economic growth is assumed to bring. At a 3 percent annual growth rate, global economic output will increase almost twenty-fold in a hundred years. Subtracting 10 percent, 30 percent, or even 50 percent for climate damages after a hundred years will still leave the world many times richer than it is today. Climate change, in short, may be bad, but even the worst seems to leave the world much better off so long as economic growth remains robust.

Instead assume that damages affect output growth *rates* rather than output *levels*. Climate change clearly affects labor productivity, especially in already hot (and poor) countries. Then the cumulative effects of damages could be much worse over time. That's the beauty—or here, the ugliness—of compound growth rates. All it took was a small but all-important change in a fundamental assumption.

Lastly, the way that climate damages are assumed to interact more generally with economic output matters a lot. DICE & Co assume that climate damages are a simple fraction of GDP: the higher the temperature, the greater that fraction. That seems like an innocuous enough assumption, but there are some stark implications. GDP and temperature just became interchangeable. Or rather: climate damages amounting to 1 percent of output can always be offset by a 1 percent increase in the output itself. *More GDP is good. If more GDP implies higher damages, increase GDP further and the world will still be better off.* It's in the DNA of many economists to make that assumption. Growth, after all, is generally good.

Alas, not all environmental damages can be offset so easily just by increasing GDP. Loss of human lives, ecosystems, or food aren't compensated so readily by increased consumer electronics. To put it in more stark terms, if the global food supply suffers from climate change, boosting GDP by building more iPhones won't do much for those who are starving. Coming up with better ways to produce food would. That's typically the rejoinder of those in favor of using the standard multiplicative model of damages. *Human ingenuity has seemingly outpaced environmental degradation in the past. Things always seem to be getting cheaper, smaller, faster, better. Technology will win the day once again.* Maybe.

But what if there are limits? What if we can't, at some point, substitute away from bad environmental outcomes in one area by increasing output further? Then more GDP will no longer compensate so easily for worse climate damages. The usual logic around economic growth being able to make up for climate damages just got turned on its head: Richer societies tend to prefer a better environment more so than poorer ones. In this world, the *higher* we can expect future GDP to be, the more valuable it is to have done something about global warming pollution today.

According to one study, if we assume that damages are additive rather than multiplicative—that food and iPhones aren't interchangeable—the "optimal" global average temperature increase is cut in half. If the standard, multiplicative version leads to around 4°C (7°F) of optimal eventual warming, making the simple change to additive damages will result in a final optimal temperature increase of below 2°C (3.6°F). That's an enormous difference, and it goes to show the importance of the assumptions that are feeding into models like DICE & Co. "Garbage in, garbage out," as the saying goes. Here it takes the form: "optimism in, optimism out." Feed a slightly different functional form through the most standard of climate-economy models, and the optimal climate policy can look very different.

|||||||||||||||||||

Once again, the inherent uncertainty is the biggest story. That goes for functional forms of the damage function as well as for lots of other factors. Even if we knew with certainty how emissions develop, how concentrations follow, how temperatures react, and how sea levels rise—and we don't—we would still need to translate it all into dollars and cents.

It's not useful to pick different types of extrapolations that deterministically project either 10 or 30 percent or more of economic damages by the time temperatures hit 6°C (11°F) above preindustrial levels. Rather, the correct approach is to do here what we just did for final temperature outcomes: look at the entire distribution of possible damages for each temperature outcome, not the expected damages conditional on any one temperature level. In other words, if temperatures were to go up 6°C (11°F), what's the probability of damages hitting 10 percent of GDP, or 30 percent, or any quantity in between or beyond? The problem is that we have no idea.

There's always a small chance that any particular final temperature wouldn't cause any damage. There's also always a small chance that it would cost the world. The most likely outcome may well be somewhere in the middle—maybe indeed somewhere in the 10 to 30 percent range for warming of 6°C (11°F)—but that's not the point. Or at least it's not sufficient. It's a "guesstimate" at best, a guess at worst.

Therefore, we simply can't give you another table the way we did for median temperature outcomes and probabilities of hitting 6°C (11°F). We don't know enough to fill in even the one row indicating average global damages at each temperature outcome. Bill Nordhaus's estimates around average expected damages—that warming of 1°C costs less than 0.5 percent of global GDP, 2°C costs around 1 percent, 4°C costs around 4 percent—could be a start. But even there, anything above around 2°C (3.6°F) is already largely guesswork. And we know much too little about the actual distribution of damages at each temperature level to estimate the third row for 50 percent or any other number for catastrophic impacts such as in table 3.2.

Table 3.2. Knowledge of economic damages decreases quickly with increased global average warming

Final temperature change	2°C (3.6°F)	2.5°C (4.5°F)	3°C (5.4°F)	3.5°C (6.3°F)	4°C (7.2°F)	4.5°C (8.1°F)	5°C (9°F)	5.5°C (10°F)	6°C (11°F)
Average global damages	1%	1.5%	2%	3%	4%	?	?	?	?*
Chance of damages >50% of economic output	?%	?%	?%	?%	?%	?%	?%	?%	?%

* Our range for average global damages from 6°C (11°F) of warming is 10 to 30 percent throughout the text, though that's hardly scientific enough to merit mention in this table. It's simply an extrapolation, using quadratic and exponential curves, from what we know—or think we know—happens at 1 or 2°C (1.8 or 3.6°F).

When it comes to high-temperature damages, the state-of-the-art economic models simply aren't much better than fitting a curve around what we know at low temperatures, and extending it into what we don't—well beyond the range of historically observed temperature increases into ones that mark uncharted territory for human civilization. Yet again, that's not to decry these modeling efforts. It's just to reiterate that inherent uncertainties will probably determine the final outcome.

||||||||||||||||||

All that begs the philosophical question: is some number better than no number at all? If Nordhaus's estimate for the average global damages caused by final warming of 6°C (11°F) is 10 percent, and a simple exponential extrapolation gives us 30 percent, should we be using this 10–30 percent range at all? And what happens if we have fundamentally mis-specified damages because, for example, they affect growth rates rather than output levels, or because damages are inherently additive rather than multiplicative?

But what's the alternative? If we didn't use these numbers in government benefit-cost analyses, we would essentially be accepting a climate damage estimate of zero. That's most definitely the wrong number. So better to go with the standard output of DICE and models like it. The U.S. government's $40 per ton figure is as good as any in that regard, even if still a likely underestimate. Let's at least run with it for now to illustrate another important point.

HOW MUCH FOR A DEGREE OF WARMING ONE HUNDRED YEARS FROM NOW?

Whether damages at 6°C (11°F) of eventual warming are 10 percent or 30 percent of global economic output or not anywhere near that range may be anyone's guess. The one thing we know for sure is that we ought to discount whichever number we get. The basic logic of discounting is sound, and ever present: It's a combination of delayed gratification and risk. Having $1 today is worth more than having it ten years from now. Answering the question of how much more seems to be as much an art as a science. But it doesn't need to be.

In fact, there's a website for it. Go to treasury.gov to find the interest rate for what are commonly viewed as the least risky investments imaginable: U.S. government bonds. Lend the United States of America $100 today for up to thirty years, and see your investment grow each year at the rate displayed there. More specifically, you'd want to look for Treasury Inflation-Protected Securities or TIPS. That way what you see is what you get in purchasing power. Inflation won't eat into your earnings. That rate has been hovering at around 2 percent a year for the past ten years. At the moment it's closer to 1 percent.

Contrast that range with the central estimate of 3 percent that the U.S. government uses in its social cost of carbon calculations. Nordhaus, in his DICE model, arrives at a default value of around 4 percent. Lord Nicholas Stern, in his *Stern Review on the Economics of Climate Change*, used 1.4 percent. He was also roundly criticized for that low choice at the time. So what is the right discount rate?

The short answer is that we don't know, but we are pretty sure the correct long-run discount rate should decline over time "toward its lowest possible value." That seems rather self-serving for anyone trying to argue for strong climate action today. A low discount rate implies that future climate damages will be more significant in today's dollars and, thus, favors strong climate action today. But there is, in fact, quite a bit of underlying science pointing in that direction. Once again, it's what we don't know that points mostly in one direction. The primary driver for low discount rates is uncertainty around the correct rate itself. Who knows what the discount rate should be a century or two from now? The less we know about the correct discount rate, the lower it ought to be. Since we know less about the right discount rate the further out we go, the rate ought to decline over time. So what exactly is that number? It's probably not 4 percent or 3 percent as currently used, but likely significantly lower, perhaps 2 percent or even less. It's so far in the future that we can't know for sure, but precautionary prudence dictates we should at least consider using low rates for long-term discounting.

Any of these numbers, though, focuses on risk-free rates. It's what you get for sure, no matter what happens to the world around you. The entire point of worrying about climate change was the unpredictability of it all. Should each potential future scenario then be discounted at the same rate?

CLIMATE FINANCE

We could do a lot worse than look to finance for cues on how to discount the uncertain future. When in doubt, ask those who actually stand to lose money from their decisions. Bob Litterman has spent most of his career at Goldman Sachs, serving in the late 1990s as head of firm-wide risk management before moving to asset management. He has lived and breathed the Capital Asset Pricing Model (CAPM) his entire life. In fact, he developed a variant, the Black-Litterman Global Asset Allocation Model. It allows for asset pricing decisions without making assumptions about expected returns for each type of asset. The less we know, the better his model performs against the standard version.

Litterman doesn't mince words when talking about the way some climate economists look at discounting: "They argue for high discount rates because of high opportunity costs for money, some estimate of the market return on capital. Waa!? If that was the sole criterion, why would anyone invest in bonds, ever? We learned in finance why this is wrong sometime in the 1960s." In fact, CAPM was developed in the 1960s, and it has one simple premise: If an investment's fortunes rise in tough economic times, it will be more valuable than an identical investment that rises and falls with the market. That link between its returns and the returns of the market is called *"beta."* Low *beta* implies a weak link. A weak link increases the value of the investment.

That, in a sense, is the only reason why anyone would invest in government bonds that pay 1 or 2 percent rather than earn an expected 7 percent in the stock market. A high overall return is good, but it's much less valuable if it pays off only in already good economic times—a high *beta*. U.S. government bonds have a low expected return but they also have a low *beta*. Many balanced investment portfolios include at least some bonds as a rainy-day fund for tough economic times.

This comes with stark implications for climate policy: If we somehow think that climate damages are small and will be worse when the economy is strong, discount rates should be higher. It will be fine to live through episodes of extreme weather events, say, because the storms won't be all that bad, and they'll hit only when GDP is high. This is one view supporting high discount rates. On the other hand, if we believe that climate damages will be large and go hand in hand with times when the economy is doing poorly, discount rates should be low. That might be a world in which climate change implies more extreme heat days, which in turn decrease labor productivity and, thus, GDP.

Or more directly, if there is around a 10 percent chance of a climate catastrophe that crashes economies and alters life as we know it, unless we change course, Finance 101 tells us that the discount rate on those damages far into the future should be low—maybe even lower than the 1 to 2 percent rate applied to assessing risk-free bonds. How low? Nobody knows for sure, but this is where we need to take a quick detour into Finance 102.

WALL STREET PUZZLES

Despite all its sophistication, modern finance leaves us with a number of fundamental puzzles. The "equity premium puzzle" tops that list. Investing in U.S. stocks returns, on average, 5 percent more than investing in U.S. government short-term bonds. This simple fact has haunted economists for decades. Standard economic models simply can't replicate these basic facts. People ought not to be so risk-averse as to warrant that large a premium for investing in risky stocks. Yet they are. What gives?

Daily stock prices are among the most well-known facts. Newspapers print them. Comprehensive databases are freely available online. So questioning the underlying data won't get us far. It's also tough to see how we could blame it on laziness, biases, or some other human quirks that may or may not contribute to the puzzle. There's a lot of money at stake, and most of it is managed by professionals, who should know better. The most natural place to look for the culprit then is economic theory itself. We know that each model simplifies reality. Do the standard models simplify too much for their own good?

It turns out that introducing potentially catastrophic risks to the standard models explains and even reverses the equity premium puzzle: fat tails in action. Market outcomes aren't defined by the average fluctuations on a typical day. They are much more defined by what happens during extreme events, the kinds of things that should never happen but have since given us at least a week's worth of "black" days in the past a hundred and fifty years: from *Black Monday* in October 1987 to *Black Friday* in September 1869 to an entire *Black Week* in October 2008. Taking these sorts of catastrophic risks more seriously justifies large equity premiums, the amounts of money investors need to be paid to take the risk.

The same holds for climate risks. Potentially catastrophic climate events demand a "risk premium." The higher the chance of these catastrophes, the more we ought to seek out the climate-equivalent of risk-free government bonds: avoiding carbon emissions in the first place.

There's one more complicating factor to this story, and it comes back to discount rates and the all-important *beta*. The reason anyone invests in government bonds is because of their low *beta*, which makes them pay off in all states of the world, including the bad. Standard asset pricing models value these investments by assigning a low, sometimes even *negative* discount rate. The latter comes into play, for example, with contrarian short sellers that earn more when the stock market is lower.

That same insurance thinking ought to apply to climate damages, or rather to avoiding them. Bob Litterman, drawing the link to climate: "If the risk premium is large enough, then the insurance benefits could even require a negative discount rate and such a high current price of emissions that the price would actually be expected to drop over time as the problem diminishes and uncertainty is resolved." The point seems blindingly obvious from an asset pricing perspective. It will come as a surprise to most focused on the proper way of discounting in the climate arena, where tying discount rates to "opportunity costs" or expected "market returns" is common practice—and where Lord Nicholas Stern's 1.4 percent has long been considered the lower bound of what is an acceptable rate. But there is nothing magical about 1.4 percent, or about 1 percent. Even 0 percent doesn't need to be a lower bound in theory. It isn't to those shorting Wall Street. If investing in a project pays more in the hardest of economic times, the proper discount rate may need to be lower than the lowest risk-free rate. We are not at all sure this is the relevant case for climate change, but it is a definite possibility, representing yet another big uncertainty.

With a large chance of catastrophe—a 10 percent chance of eventually hitting 6°C (11°F), say—and with that catastrophe associated with large economic costs—10 or 30 percent (or even much more) of global economic output—the proper treatment of these climate damages is to discount them at a rate maybe even lower than the risk-free rate of government bonds. As always, it's tough to pick a single number, but it makes it harder to argue for discount rates much above 1 or 2 percent.

TIMING IS EVERYTHING

We keep saying "eventual" in connection with warming of 6°C (11°F) and other such extreme scenarios, because any of these catastrophic temperature increases would play out over many decades and centuries. Large global average temperature increases won't happen tomorrow; nor would catastrophe strike overnight, at least not based on this calculation. In fact, the higher the final temperature increase, and the higher the chance of ultimate catastrophe, the longer it will take for both to materialize. That points to one of the more profound characteristics of climate change: its long-term nature. But it clearly doesn't mean that we can relax for the time being.

If a civilization-as-we-know-it-altering asteroid were hurtling toward Earth, scheduled to hit a decade hence, and it had, say, a 5 percent chance of striking the planet, we would surely pull out all the stops to try to deflect its path.

If we knew that same asteroid were hurtling toward Earth a century hence, we may spend a few more years arguing about the precise course of action, but here's what we wouldn't do: We wouldn't say that we should be able to solve the problem in at most a decade, so we can just sit back and relax for another 90 years. Nor would we try to bank on the fact that technologies will be that much better in 90 years, so we can probably do nothing for 91 or 92 years and we'd still be fine.

We'd act, and soon. Never mind that technologies will be getting better in the next 90 years, and never mind that we may find out more about the asteroid's precise path over the next 90 years that may be able to tell us that the chance of it hitting Earth is "only" 4 percent rather than the 5 percent we had assumed all along. That last point—increased certainty around the final impacts—is precisely where climate change has proven so vexing. Our estimate of the range of climate sensitivity isn't any more precise today than it was over three decades ago. And the chance of eventual climate catastrophe isn't 5 percent; our rough calculation based on IEA projections shows that it's likely closer to 10 percent or even more.

WHAT'S YOUR NUMBER?

Climate change is beset with deep-seated uncertainties on top of deep-seated uncertainties on top of still more deep-seated uncertainties. And that's just for going from emissions to concentrations to final temperatures. Further uncertainties prevent us from simply translating temperatures into economic damages, and none of that yet clarifies the uncertainties around the correct discount rates to calculate the optimal carbon price today. In each of these steps, though, one thing is clear: because the extreme downside is so threatening, the burden of proof ought to be on those who argue that fat tails don't matter, that possible damages are low, and that discount rates ought to be high.

As little as we know about many of these uncertainties, we do know that the chance of eventual catastrophic warming of 6°C (11°F) or more isn't zero. It's slightly greater than around 10 percent under our conservative calibration.

Associated damages are anyone's guess, but we can only consider the implied "guess" of 10 percent of global economic output ventured by Bill Nordhaus's DICE model a lower bound. Following the same, admittedly imperfect logic could yield estimates anywhere from 10 percent to 30 percent or even well beyond. We don't know where in that range the true number is. We are pretty sure it's not less than 10 percent, and we do know that no one else knows the true number either. The most relevant question isn't whether expected damages at 6°C (11°F) are 10 or 30 percent of global economic output. The question ought to be: what is the full distribution of damages and what is the chance of significant economic collapse?

That leaves discounting, where at least we know that looking for an expected market return on capital to arrive at a discount rate of, say, 4 percent may be turning a blind eye on decades of asset pricing theory and practice. If we omit the rosy scenario where climate damages are small and will be worse when the economy is strong, we are looking at much lower rates than are currently bandied about. We don't know whether the right rate should be 2 percent, 1 percent, or even below. There may not be a single *climate beta*—the link between climate damages and the general health of the economy—to justify using any one particular rate. But we can be pretty sure that the presence of big uncertainties around high final temperatures and catastrophic damages should drive discount rates down, not up. A rate of 2 percent might be our estimate for damages fifty years hence, and whatever rate it is, it ought to decline over time.

Where does all of that leave us? First, with the realization that it's easy to criticize. It's tougher to come up with a constructive alternative. Table 3.2, showing actual climate damages, is mostly blank for a reason, and it's not for lack of trying.

If the question is what single number to use as the optimal price of each ton of carbon dioxide pollution today, the answer should be: at least $40 per ton of carbon dioxide, the U.S. government's current value. We know it's imperfect. We are pretty sure it's an underestimate; we are confident it's not an overestimate. It's also all we have. (And it's a lot higher than the prevailing price in most places that do have a carbon price right now—from California to the European Union. The sole exception is Sweden, where the price is upward of $150. And even there, key industrial sectors are exempt.)

If the next question is how to decide on the proper climate policy, the answer is more complex than our rough benefit-cost analysis suggests. Pricing carbon at $40 a ton is a start, but it's only that. Any benefit-cost analysis relies on a number of assumptions—perhaps too many—to truly come up with one single dollar estimate based on one representative model of something as large and uncertain as climate change.

Since we know that fat tails can dominate the final outcome, the decision criterion ought to focus on avoiding the possibility of these kinds of catastrophic damages in the first place. Some call it a "precautionary principle"—better safe than sorry. Others call it a variant of "Pascal's Wager"—why risk it, if the punishment is eternal damnation? We call it a "Dismal Dilemma": while fat tails can dominate the analysis, how can we know the relevant probabilities of rare extreme scenarios that we have not previously observed and whose dynamics we understand only crudely at best? The true numbers are largely unknown and may simply be unknowable.

|||||||||||||||||

In the end, it's risk management—existential risk management. And it comes with an ethical component. Precaution is a prudent stance when uncertainties about catastrophic risks are as dominant as they are here. Benefit-cost analysis is important, but it alone may be inadequate, simply because of the fuzziness involved with analyzing high-temperature impacts.

Climate change belongs to a rare category of situations where it's extraordinarily difficult to put meaningful bounds on the extent of possible planetary damages. Focusing on getting precise estimates of the damages associated with eventual global average warming of 4°C (7°F), 5°C (9°F), or 6°C (11°F) misses the point. The appropriate price on carbon is one that will make us comfortable enough to know that we will *never* get to anything close to 6°C (11°F) and certain eventual catastrophe. *Never*, of course, is a strong word, since we know the chance of any of these temperatures happening even based on today's atmospheric concentrations can't be brought to zero.

One thing we know for sure is that a greater than 10 percent chance of eventual warming of 6°C (11°F) or more—the end of the human adventure on this planet as we now know it—is too high. And that's the path the planet is on at the moment. With the immense longevity of atmospheric carbon dioxide, "wait and see" would amount to nothing other than willful blindness.

IN THE
OPINION OF TWO ECONOMISTS MUSING ON:

WILLFUL BLINDNESS

Chapter 04

SEVEN BILLION PEOPLE AND UNTOLD FUTURE GENERATIONS *V.*
THOSE STANDING IN THE WAY OF SENSIBLE CLIMATE ACTION
On Writ of Certiorari to the Court of Public Opinion

The willful blindness doctrine is old news in criminal defense cases. Just because you turned your back when your partner pulled the gun on the bank teller doesn't mean you won't be held accountable for aiding in the robbery.

That doctrine has since reached beyond the sphere of criminal law. The U.S. Supreme Court, for example, has applied the principle to patents. It took up a case about an "innovative deep fryer": *Global-Tech Appliances, Inc., et al. v. SEB S.A.*, 563 No. 10–6 (May 31, 2011). In the Supreme Court's words: "[P]ersons who know enough to blind themselves to direct proof of critical facts in effect have actual knowledge of those facts." More specifically, the Court further adds two basic requirements for willful blindness that "all agree on": "First, the defendant must subjectively believe that there is a high probability that a fact exists. Second, the defendant must take deliberate actions to avoid learning of that fact." *Ibid.*

The key words are "high probability" and "deliberate action." Little is ever certain. Demonstrating willful blindness requires showing that the defendant must have been knowledgeable enough to realize—with high probability—that something's happening. And then, said defendant must have taken deliberate actions to avoid acting on that knowledge.

Shift from criminal defense to climate change, and from Supreme Court legalese to the colloquial interpretation of what it means to be "willfully blind." Climate change is bad. Not acting on it makes it worse. After decades of science and years of public discussion saying as much, there's no other way than to call those opposing—or outright denying—this reality "willfully blind." Some may simply be "blind." But for many, motivated reasoning leads to conclusions in direct conflict with science.

It's tempting to stop there. The debate around *whether* to act should be over. To some extent, even the debate around *how* to act is over. Yes, there are some academic and also practical disagreements around what to do—whether to tax or cap carbon, how to put either into practice, or how to approximate a price on carbon using other policies like fuel-economy standards for cars or carbon pollution standards for power plants. Some of these policies—like performance standards—may well have merit on their own, but they do not add up to enough. The ultimate goal is clear: price carbon. None of this is a secret. Dare we say that anyone who pretends otherwise is, once again, willfully blind to reality?

The question we are left with is: how high should that price on carbon be? This is where there is room for actual discussion, academic and otherwise. To be clear, that discussion cannot—must not—excuse inaction now. We do know for a fact that the prevailing current price of close to zero in most countries, with some notable exceptions, is much too low. The central estimate of the U.S. government's comprehensive review of the "social cost of carbon" is a price of around $40 per ton of carbon dioxide.

That $40 per ton, however, can be only the starting point. Most of what we know about the science seems to point to the fact that the number should be higher. Most everything we don't know pushes it higher still. A serious look at the host of uncertainties involved in calculations that arrive at the $40 figure makes that clear. The past chapter's "fat tails" may well dominate all else. So how high should it go?

LESS THAN INFINITY—INFINITELY LESS

If the risk of catastrophe is sufficiently high, and the catastrophe itself is sufficiently bad, it's tempting to conclude that we ought to avoid it at all cost. Don't stop at $40, or $400, or even at $4,000 per ton of carbon dioxide. If the catastrophe is infinitely bad, the optimal cost of each ton of carbon dioxide pollution beyond any thresholds that triggers the catastrophe would be infinity as well. This is true even if there's, say, "only" a 10 percent chance of catastrophe. Infinity times any number is still infinity. The precise mathematical verdict then would be to spend all the money in the world to avoid that outcome. Or in practical terms: ban the intentional release of carbon dioxide—the burning of coal, oil, and gas as well as deforestation—altogether. Stop driving cars with an internal combustion engine. Ground all commercial planes. Turn off fossil-fueled power plants. Stop modern life as we know it.

That can't be the right policy prescription. For one, even if we managed to decrease the chance of climate catastrophe from 10 percent to perhaps 1 percent, there's still an infinite cost to be avoided: 1 percent times infinity is still infinity. Standard benefit-cost analysis falls apart as soon as we introduce the term "catastrophe" and describe it as infinitely costly.

Little, of course, is ever truly infinitely costly. Not even death is. The "value of a statistical life" may appear as heartless a calculation as the name implies, but the logic is unassailable. When making small, everyday decisions like whether to wear a seatbelt, or bigger ones like which job to get, people don't value their lives infinitely much. If your chance of dying on the job in one profession versus another is twice as high, wages may be appropriately higher, but not infinitely so. It may be a bit of a stretch to apply the value of an individual statistical life to a planetary catastrophe, but the analogy holds.

It's not hard to see that eventual global warming of 6°C (11°F) would be nothing short of catastrophic, destroying nature and civilization as we now know them. That doesn't yet mean we should throw infinite amounts of money at solving the problem. We ought to find a sensible balance between overreaction and inexcusable inaction.

One possible guide is to look at the risk of catastrophe itself. Shortly after 9/11, Vice President Dick Cheney postulated that, "if there's a one percent chance that Pakistani scientists are helping [Al Qaeda] build or develop a nuclear weapon, we have to treat it as a certainty in terms of our response." That, in fact, is a false equivalency. One percent isn't certainty. Instead, one ought to consider as a crucial metric the probability of the event itself. An existential risk with a truly tiny chance of occurring deserves less of our attention than one with a 10 percent or even 1 percent probability.

We would be lucky if the probability of existential climate risk were only 1 percent, as in Mr. Cheney's Al Qaeda scenario. We have seen with our own rather conservative calculation that the risk of having eventual average global temperatures go up by more than 6°C (11°F) is about 10 percent, ten times higher.

By now we are running up against another possible constraint: Climate change isn't the only potential catastrophe hitting the planet. What if an asteroid struck the planet and wiped out civilization long before the worst effects of climate change began to show? Or what about a pandemic? Or the chance of nuclear terrorism? Or biotechnology, nanotechnology, or robots gone amok? How should we allocate limited amounts of money to each of these existential risks?

WORST-CASE SCENARIOS

Opinions differ on what should rightly be called an "existential risk" or planetary-scale "catastrophe." Some include nuclear accidents or terrorism. Others insist only nuclear war or at least a large-scale nuclear attack reaches dimensions worthy of the "global" label. There are half a dozen other candidates that seem to make it on various lists of the worst of the worst. As we shall soon see, it's tough to come up with a clear order. In addition to climate change, let's consider, in alphabetical order, asteroids, biotechnology, nanotechnology, nukes, pandemics, robots, and "strangelets."

That might strike some as a rather short list. Aren't there hundreds or thousands of potential risks? One could imagine seemingly countless ways to die in a traffic accident alone. That's surely the case. But there's an important difference. While traffic deaths are tragic on an individual level, they are hardly catastrophic as a class.

Every entry on our list has the potential to wipe out civilization as we know it. All of the worst-case scenarios are global. All are highly impactful and mostly irreversible in human timescales. Most are highly uncertain.

Only two—asteroids and climate change—allow us to point to history as evidence of the enormity of the problem. For asteroids, go back 65 million years to the one that wiped out the dinosaurs. For climate, go back a bit over three million years to find today's concentrations of carbon dioxide in the atmosphere and sea levels up to 20 meters (66 feet) higher than today.

In the final analysis, climate change is far from the only potential catastrophe humanity ought to be worrying about. Others, too, deserve more attention, and funding. Though that doesn't apply to all.

Strangelets are something straight out of science fiction: stable strange matter with the potential of swallowing the Earth in a fraction of a second. They have never been observed. It may be theoretically impossible. If it is possible, though, there may be a chance that large heavy-ion colliders like CERN could create them. That has prompted research teams to calculate the likelihood of a strangelet actually happening. Their verdict: negligible. Concrete numbers hover between 0.0000002 percent and 0.002 percent. That's not zero, but it might as well be. It's certainly nowhere near the 10 percent chance of ultimate climate catastrophe on our current path.

So yes, swallowing the entire planet would be the ultimate bad—clearly worse, say, than melting the poles and raising sea levels by several meters or feet. Stranger things have happened. But strangelets very, very, very likely won't.

Probabilities matter. Those problems that are truly unlikely should not command a lot of society's attention. Quantum physics tells us there's an infinitesimally small chance of our planet jumping off its current path around the sun and shooting out into space. It's most definitely not a possibility humanity should spend any time worrying about. Strangelets are slightly different in that they could, in fact, be a human creation. But that still doesn't mean they deserve society's attention. The chance is simply too small.

If we could rank worst-case scenarios by how likely they are to occur, we'd have taken a huge step forward. If the chance of a strangelet is so small as to be ignorable, probabilities alone might point to where to focus. But that's not all. The size of the impact matters, too. So does the potential to respond.

Asteroids come in various shapes and sizes. We began this book by looking at the one that exploded above Chelyabinsk Oblast in February 2013. The impact injured 1,500 and caused some limited damage to buildings. We shouldn't wish for more of these impacts to happen just for the spectacular footage, but we'd be hard-pressed to call an asteroid of that size a "worst-case scenario." It's not. NASA's attempts at cataloguing and defending against objects from space aims at much larger asteroids, the ones that come in civilization-destroying sizes. Astronomers may have been underestimating the likelihood of Chelyabinsk Oblast–sized asteroids all along. That's a problem that needs to be rectified, but it's not a problem that will wipe out civilization. If we estimated the likelihood of a much larger impact incorrectly, the consequences could be significantly more painful.

Luckily, when it comes to asteroids, there's another feature working for us. Science should be able to observe, catalogue, and divert every last one of these large asteroids—if sufficient resources are provided. That's a big *if*, but not an insurmountable one: a National Academy study puts the cost at $2 to 3 billion and ten years' research to launch an actual test of an asteroid deflection technology. That's much more than we are spending at the moment, but the decision seems rather easy: spend the money; solve the problem; move on.

||||||||||||||||||

Now we are down to biotechnology, nanotechnology, nukes, pandemics, and robots as contenders against which climate change needs to be measured. Or does it?

One response to any list like that would be to say that each such problem deserves our (appropriate) attention, independently of what we do with any of the others. If there's more than one existential risk facing the planet, we ought to consider and address each in turn. A typical homeowner insurance package will protect against fire damage. If your home sits near a fault line, you may also want to buy earthquake insurance. Those with additional flood risk buy flood insurance as well. And so on. The same should go for catastrophe policy.

That logic has its limits. If catastrophe policies were to eat up all the resources we have, we'd clearly have to pick and choose. But we don't seem to be anywhere close. A first step then should always be to turn to benefit-cost analysis, which in turn is something that every U.S. president since Ronald Reagan has affirmed as a guiding principle of government policy.

Ideally, society should conduct serious benefit-cost analyses for each (remaining) worst-case scenario: estimate probabilities and possible impacts, multiply the two, and compare it to the costs of action in each instance. If climate change *and* biotechnology *and* nanotechnology *and* nukes *and* pandemics *and* robots emerge as problems worthy

of more of our attention, society should devote more resources to each.

But we can't just hide behind standard benefit-cost
analysis that ignores extremes. Each of these scenarios may
also have their own variant of fat tails: underestimated and
possibly unquantifiable extreme events that could dwarf
all else. The analysis soon moves toward some version of
a precautionary principle focused on extreme events. The
further we move away from standard benefit-cost analysis,
the more acute then the need to compare across worst-case
scenarios.

That comparison is getting increasingly difficult. We
cannot dismiss out of hand any of the five remaining
worst-case scenarios. Their probabilities aren't so close to
zero as to be negligible. The potential downsides are large.
Ask anyone working on nuclear nonproliferation, and
they might well argue that nuclear terrorism is worse than
climate change. Ask a virologist, and they'll tell you that
society is inadequately prepared to combat pandemics.

What then, if anything, still distinguishes climate change
from the five remaining others?

For one, it's the relatively high chance of eventual planetary catastrophe. Our own analysis from the last chapter
puts the likelihood at around 10 percent, and that's for an
indisputable global catastrophe. Climate change would
trigger plenty of catastrophic events with temperatures
eventually rising by much less than 6°C (11°F). Many scientists would name 2°C (3.6°F) as the threshold, and we
are well on our way to passing and exceeding that, unless
there is a major global course correction.

Second, the gap between our current efforts and what's
needed on climate change is enormous. We are no experts

on any of the other worst-case scenarios, but there at least it seems like lots is already being done. Take nuclear terrorism. The United States alone spends many hundreds of billions of dollars each year on its military, intelligence, and security services. That doesn't stamp out the chance of terrorism. Some of the money spent may even be fueling it, and there are surely ways to approach the problem more strategically at times, but at least the overall mission is to protect the United States and its citizens. It would be hard to argue that U.S. climate policy today benefits from anything close to this type of effort. As for mitigating pandemics, more could surely be spent on research, monitoring, and rapid response, but here too it seems like needed additional efforts would plausibly amount to a small fraction of national income.

Third, climate change has firm historical precedence. Humans have never experienced it, but the planet has. Some of the other potential global catastrophes often rely on a heavy dose of science fiction. Autonomous robots reproducing and taking over the world may be the most extreme example. Not that it can't ever happen, but it certainly hasn't happened before. Climate change has.

There's ample reason to believe that pumping carbon dioxide into the atmosphere is reliving the past—the distant past, but the past nonetheless. The planet has seen today's carbon dioxide levels before: over three million years ago, with sea levels some 20 meters (66 feet) higher than today, and camels roaming the high Arctic. There are considerable uncertainties in all of this, but there's little reason to believe that humanity can cheat basic physics and chemistry. Many of the effects from climate change are unprecedented in human timescales, but that doesn't make them unprecedented in geological time scales: no need for science fiction to tell the story.

Contrast the historical precedent of climate change with that of biotechnology, or rather the lack of it. The fear that bioengineered genes and genetically modified organisms (GMOs) will wreak havoc in the wild is a prime example. They may act like invasive species in some areas, but a global takeover seems unlikely, to say the least. Much like climate change, historical precedent can give us some guidance. But unlike climate change, that same historical precedent gives us quite a bit of comfort. Nature itself has tried for millions of years to create countless combinations of mutated DNA and genes. The process of natural selection all but guarantees that only a tiny fraction of the very fittest permutations has survived. Genetically modified crops grow bigger, stronger, and are otherwise pesticide-resistant. But they can't outgrow natural selection entirely. None of that yet guarantees that scientists wouldn't be able to develop permutations that could wreak havoc in the wild, but historical experience would tell us that the chance is indeed slim.

Reassuringly, the best scientists working on biotechnology seem to be much less concerned about the dangers of "Frankenfoods" and GMOs than the general public. The reverse holds true for climate change. The best climate scientists appear to be significantly more concerned about ultimate climate impacts than the majority of the general public and many policy makers.

Some of these same climate scientists—knowing what they know about the science, and knowing what they know about human responses to the climate problem—have seemingly moved on. And they haven't moved on to analyzing any of the other worst-case scenarios, believing that climate isn't all that bad. Quite the opposite: Some have moved on to looking for solutions to the climate crisis in an entirely different realm, searching for anything that could pull the planet back from the brink of a looming catastrophe. Their focus: geoengineering.

CHAPTER 5

Bailing Out the Planet

IN JUNE 1991 AND WITH A YEAR TO GO, preparations for the Rio Earth Summit were in full swing. "Sustainable development" was in vogue. Who could disagree that humanity ought to "make development sustainable to ensure that it meets the needs of the present without compromising the ability of future generations to meet their own needs"?

The excitement was palpable. It might still be possible to achieve sustainable development "by the year 2000 and beyond," as the General Assembly of the United Nations had called for. There was only one problem: the Earth's atmosphere had already warmed by more than 0.5°C (0.9°F) since the industrial revolution, with all trends pointing higher still.

China had just emerged from a couple decades of market-based economic reforms and was on the cusp of pulling hundreds of millions of its citizens out of abject poverty. The best technologies available at the time meant that China would spend the next decade largely duplicating what the United States, Europe, and others had done before to support their comfortable status as the world's rich countries: Burn coal, oil, and natural gas—mostly coal—and dump the resulting carbon dioxide into the air, further heating the planet. There was only so much President George H. W. Bush could do by signing the 1992 Earth Summit declaration "Agenda 21," other than give heartburn and a rallying cry to future generations of right-wing conspiracy theorists. But all that was still a year out. President Bush and over a hundred fellow heads of state would not fly to Rio until June 1992.

Meanwhile, Mount Pinatubo, a volcano that had been dormant for over four hundred years, had begun to rumble by April 2, 1991. Soon thereafter, Filipino authorities gave their first evacuation orders. Two months later, volcanic activity went into overdrive, culminating in a final explosion on June 15. Its ash, rocks, and lava buried the surrounding area. To make things worse, Typhoon Yunya slammed the area that very same day. The resulting floods combined with the effects of the explosion displaced over 200,000 Filipinos. Over 300 died.

The costs were real. So were the benefits: As a direct result of the volcanic explosion, global temperatures temporarily decreased by about 0.5°C (0.9°F), wiping out the entire temperature effects of human-caused global warming up to that point. The reduction in temperatures hit its peak just around the time of the Rio Earth summit a year later.

Mount Pinatubo did all that by putting around 20 million tons of sulfur dioxide into the stratosphere. That relatively small amount counteracted the global warming effect of around 585 billion tons of carbon dioxide that humans had managed to accumulate in the atmosphere by then. (By now, twenty years later, the total number accumulated in the atmosphere is around 940 billion tons, and all signs are still pointing up.)

The leverage ratio of sulfur to carbon dioxide in geo-engineering terms is enormous. The sulfur dioxide released by Mount Pinatubo reduced temperatures by about 30,000 times as much as the same amount of carbon dioxide would have increased them.

It's tempting to draw the link to nuclear technology. Little Boy, the atomic bomb dropped over Hiroshima, had roughly 5,000 times as much power as traditional explosives with the same amount of materials.

The comparison to nuclear technology also shows the possible path ahead. The Titan II missile was developed just 15 years after Little Boy was dropped. It contained more power in its warhead than all the bombs dropped in World War II combined, *including* Little Boy. If geoengineering technology advanced even a fraction as quickly, it's hard to imagine the technologies that could become available to counteract atmospheric carbon dioxide. Even using today's technology, a more targeted geoengineering intervention could possibly achieve leverage ratios near a million to one.

The similarities to the leverage of nuclear bombs are striking. There's also an important difference: Both nuclear and conventional explosives destroy, whereas geoengineering *counteracts* carbon dioxide. At least in principle, the enormous leverage has the potential to do immense good.

THE PROMISE AND PROBLEMS
OF GEOENGINEERING

Without considering the very real costs and lives lost, Mount Pinatubo's effect on global temperature was presumably a good thing. If we could wipe out two centuries of accumulated, human-caused global warming by turning a knob, why not go for it?

There are a few problems with that simple picture. Mount Pinatubo decreased the *indirect*, if all-too-real, effects of carbon dioxide in the atmosphere: The 20 million tons of sulfur dioxide created a sunshade that dimmed the amount of radiation from the sun by about 2 to 3 percent throughout the following year. The eruption did nothing to counteract the *direct* effects of carbon pollution, like oceans turning more acidic after absorbing some of the added carbon dioxide. You can't expect one volcanic eruption to solve everything. Then again, Mount Pinatubo didn't just fail to solve problems, it created more.

As much as participants in the 1992 Earth Summit could have been elated by the decreased global average warming, they must have been distraught by the accompanying low levels of stratospheric ozone. Combine the volcano's sulfur dioxide and other gunk with certain types of pollution we humans send into the stratosphere, and you may get stratospheric ozone depletion of the type that gave us the ozone hole over the South Pole—but now the ozone depletion would occur over the tropics as well.

If that wasn't enough, Mount Pinatubo is also invariably blamed for flooding along the Mississippi river in 1993 and for droughts elsewhere. The volcanic eruption coincides with the beginning of a remarkably global dry spell lasting about a year. Direct links are difficult to establish, but that only makes it more problematic. If we could draw a direct line from Mount Pinatubo to sub-Saharan African droughts, we'd at least know whom to hold responsible. Without that link, speculation runs rampant.

||||||||||||||||||

What if instead of a volcano it had been a group of scientists launching an experiment to counteract two centuries of global warming just in time for the Rio Earth Summit?

One would hope, at a minimum, that the experiment could have been designed in a way to avoid the 200,000 evacuations and 300 deaths. But even without those all-too-direct effects, it would have been hard to imagine a university's Institutional Review Board, the group charged with overseeing the safety of research activity, approving the experiment. It's often hard enough to get approval for a simple e-mail survey, asking test subjects to deploy their computer mice and answer a few benign questions, not to say anything of injecting patients with new drugs that hold promise but may well have negative side effects. Now imagine intentionally injecting the stratosphere with tiny custom-designed particles to mimic the effects of Mount Pinatubo, with the expressed purpose of altering the global climate.

Forget Institutional Review Boards. The public may have a word or two to say here—as it should. Even if the only effect of the experiment would be to lower global temperatures perfectly uniformly with no regional difference whatsoever (something that turns out not to be the case), it would still be hard to agree on the "right" temperature.

If you live in Cape Town, San Francisco, or along the Mediterranean, you pretty much enjoy the most stable, ideal climates anywhere on Earth. Why change that?

If you live at higher latitudes, a few degrees of warming may not be all that bad for you personally. Why dial that back?

And if we do dial it back, where should we stop? Pre-industrial levels seem like a reasonable target. But today seems fine, too.

There is no *right* answer to any of these questions, other than to say that we would need strong, global institutions and well-formed governance processes to make these decisions in a way that considers a breadth of voices in the most democratic and well-informed way possible. That's a lot to ask. We don't have a global government. Instead, we need to work with what we have. That's a fragmented global governance complex with imperfect representation and even more imperfect decision processes. Decision making in Washington, D.C., may be at a standstill, but at least there is a formal process to make the decisions. On a global level, we have yet to create the institutions that allow us to even have the conversation.

Fortunately, we are still far from having to make decisions around deploying geoengineering to cool the planet. Unfortunately, the failure to harness market forces to deal with global warming pushes us relentlessly in that direction, whether we want it or not.

FREE RIDERS, MEET THE *FREE DRIVERS* OF GLOBAL WARMING

Climate change is a problem because too few of us consider it one. And those of us who do consider it a problem, or worse, can do little about it unless we get everyone else to act. Either we solve this problem for everyone, or we solve it for none of us.

That, in a nutshell, is what makes climate change so difficult to solve. You alone can do little beyond scream to get the right policies in place, which could then guide the rest of us in the right direction. Meanwhile, the overwhelming majority of the seven billion of us on this planet are *free riders*. We enjoy the going while the going is good. We don't pay for the full cost of our actions.

Worse, polluting is subsidized worldwide to the tune of some $500 billion dollars per year. That averages out to a subsidy of around $15 per ton of carbon dioxide, much of it in oil-rich and developing countries like Venezuela,

Saudi Arabia, and Nigeria as well as China, India, and Indonesia. Every one of these dollars is a step away from setting the right incentives. Instead of paying for the privilege of polluting, we are paid to pollute. (Meanwhile, carbon dioxide prices in most of the United States, with the notable exception of California, are about zero, or close to it. That estimate assumes subsidies of around $3 per ton of carbon dioxide roughly balanced by direct and indirect measures such as efficiency standards and renewables mandates.)

Every time you board a plane to fly from New York to San Francisco and back, you put roughly a ton of carbon dioxide into the air, some of which will stay there for decades or even centuries after your trip. That's you personally, not the whole plane, which emits a couple hundred times more. And that ton will cause at least around $40 worth of damages to the economy, ecosystems, and health.

Assume, for argument's sake, that all seven billion of us board a plane every given year. Also assume that each flight causes about one ton of carbon dioxide pollution. (The latter is approximately right for transatlantic flights from Europe to the United States. The former isn't true at all. Flying, like most other sources of global warming pollution, is largely a rich people's activity. There are about 30 million commercial flights per year globally, transporting three billion passengers. That doesn't mean three billion different people fly each year. It's far fewer than a billion who take several flights each year. But let's stick with the seven billion passengers for now.)

If seven billion of us flew, and each of us caused one additional ton of carbon dioxide pollution, we'd collectively cause seven billion times $40 worth of damages. Divided by seven billion, we'd get back to each person facing a price of $40.

Everyone ends up paying that $40. But no one is facing the *right* $40.

That's the crux of the problem. Instead of having *your* $40 worth of damages added to *your* ticket, you pay a fraction of a fraction of a penny for the damages caused by everyone else's flights. The same holds true for everyone else. Every person faces the exact same choice set: "my benefit, seven billion people's cost." We all collectively bear the cost of pollution, but no one faces *their own* global warming pollution costs of travel. As a result, we fly too much and saddle society with enormous costs: seven billion times $40, to be exact.

Total costs are large. But no one has the right incentives to try to do something about them. On an individual basis, the damages your flying causes—the $40—goes back to the fraction of a fraction of a penny for each of the other seven billion. No one will be motivated to rise up and try to prevent you from boarding that plane, or at least to make you pay for the damage your flight will cause. Voluntary coordination is out. Getting seven people to agree on anything is tough; getting seven billion to agree is impossible. That's where governments need to come in, and even there we find global cooperation very difficult.

So far, not so good. But *free riding* is only half the problem.

Free driving may be just as important. That's where geo-engineering gets behind the wheel, and we end up back at Mount Pinatubo. About 20 *million* tons of sulfur dioxide managed to wipe out the global warming effects of 585 *billion* tons of carbon dioxide in the atmosphere. That's leverage. It's also another way of saying that it would probably be cheap, if scientists were able to duplicate the effects of Mount Pinatubo intentionally—"cheap," that is, in the narrow sense of the direct engineering costs of transporting

20 million tons of material into the stratosphere, not necessarily cheap when looking at the full consequences.

IIIIIIIIIIIIIIII

We may hate the idea of countering amazing amounts of pollution with yet more pollution of a different type. But the entire enterprise is simply too cheap to ignore.

And it's not like anyone would literally do as Mount Pinatubo did and dump 20 million tons of sulfur dioxide into the stratosphere. At the very least, given current technology and knowledge, the sulfur would likely be delivered in the form of sulfuric acid vapor. Sooner rather than later, we may be looking at particles specifically engineered to reflect as much solar radiation back into space with as little material as possible. That would mean less material to achieve the same impact. It may be a fleet of a few dozen planes flying around the clock. Some have gone as far as to calculate how many commercially available Gulfstream G650 jets it would take to haul the necessary materials. The specifics are indeed too specific. What matters is that the total costs are low, both compared to the damage carbon dioxide causes and the cost of avoiding that damage by reducing emissions.

Actual numbers are all over the place, and all of them are based on estimates, but most put the direct engineering costs on the order of $1 to 10 billion a year. Those are the engineering costs of getting temperatures back down to preindustrial levels. It's not nothing, but it's well within the reach of many countries and maybe even the odd billionaire.

If a ton of carbon dioxide emitted today costs $40 over its lifetime, we are talking pennies per equivalent ton. That's three orders of magnitude lower, and it's the exact parallel situation to the *free-rider* problem that has caused the problem in the first place. Instead of one person enjoying all the benefits of that cross-country round-trip and the other seven billion paying fractions of a penny each for the climate damages that one ton of carbon dioxide causes, now it's one person or (more likely) one country being able to pay for the costs of geoengineering the entire planet—all potentially without consulting the other seven billion people.

Welcome to the *free-driver* problem.

If climate change is the *mother of all externalities*, as economists like to call it, geoengineering is the *father of all externalities*. The world is the child stuck in the middle. If mom says "no," go to dad and see whether he says "yes." The chance is pretty good, seeing as he's facing the exact opposite incentives from mom: a game of good cop/bad cop on a planetary scale.

Geoengineering is too cheap to dismiss as a fringe cause developed by sinister scientists looking for the next big sensational issue to attract some attention and grant money, as some pundits would have it. If anything, it's the most experienced of climate scientists who take the issue most seriously. And not because they like to.

OF SEAT BELTS AND SPEED LIMITS

In February 1975, a who's who in biomedical research at the time descended on a small seaside resort in Pacific Grove, California, to discuss safe laboratory standards for the burgeoning discipline of recombinant DNA research. There was lots of promise but also significant danger—not least that the science would get ahead of the public and evoke a backlash in public opinion that could result in protests, defunded labs, and shuttered science programs. By all accounts, the meeting was a success. Research had, in fact, been halted ahead of the meeting because of public outcries over its possible dangers. Since then, recombinant DNA research has given us, among many other things, the hepatitis B vaccine, new forms of insulin, gene therapy, and a Nobel Prize in chemistry for Paul Berg, the co-organizer of the 1975 meeting.

The meeting also provided an example of how scientists can and should engage the public when their research hits particularly touchy subjects. Ahead of the 1975 gathering at the Asilomar Conference Facilities in Pacific Grove, even Berg's own coinvestigators had asked him to stop his research because of fears of biohazards that could lead to cancer in lab technicians or worse. The "Asilomar Process" assured scientists and helped guide science policy for decades to come.

It's almost comical to believe nowadays that a single meeting like that between a few dozen biologists, a handful of physicians, and the odd lawyer could assuage the public and policy makers alike in order to do what's right for science. You can already imagine the conspiracy theories swarming around. The newspaper and editorial headlines practically write themselves:

"How Far Is Too Far? Should Scientists Decide Their Own Limits?"
"The Brave New World of Hacking Your Genes."
"Hacking the Planet: Who Decides?"

The last of these headlines was, in fact, real. The *New Scientist* used it for an article entitled "Asilomar 2.0." That's at least how the organizers wanted it to be known. In March 2010, prominent climate scientists, budding geoengineers, a few journalists, and the odd diplomat and environmentalist descended on the Asilomar facilities to try to rekindle the spirit of 1975. It was a gathering of the who's who in another burgeoning area of research with a lot of promise and quite a bit of potential for public backlash: geoengineering.

The opening line from a co-organizer set the tone of the conference: "Many of us wished we wouldn't be here." Most scientists wished instead that the world had heeded their advice and done something about global warming pollution decades ago. The late Steve Schneider spoke passionately about his climate research that had raised some of the first alarms, going back even before 1975. He had just written his own firsthand account, *Science as a Contact Sport: Inside the Battle to Save Earth's Climate*. And he wasn't there to sell or sign books. He came to lament the fact that it had come to this. Every scientist who spoke prefaced his or her words by saying that the "told you so's" were bittersweet. The final statement from the meeting began with an unambiguously "strong commitment to mitigation of greenhouse gas emissions," addressing the root of the problem in the first place.

That's where we are now. Some of the most serious of climate scientists are looking toward geoengineering as an option—not because they like to, but because it may well be our only hope for avoiding a climate catastrophe. Mount Pinatubo–style remedies have gotten significant attention of late for precisely that reason.

These very scientists also highlight one of the key problems that comes up when discussing geoengineering. As we're sucked into the *free-driver* problem, we inevitably spend less time trying to solve the *free-rider* one. Life comes with trade-offs. Spend the better part of your workday worrying about shooting tiny sulfur-based particles into the atmosphere, and you don't spend that time worrying about getting carbon out. That's the very real trade-off faced by scientists.

The same conundrum holds outside the lab: why reduce emissions, if we know that the latest technological advance can solve the problem without changing our ways? The clearest response is simply that geoengineering doesn't actually solve the problem. It may be able to treat some of the symptoms. Pick your favorite analogy. It's like "chemotherapy" or a "tracheostomy" for the planet: a last-ditch effort to do what prevention and any other kind of treatment failed to accomplish.

For an analogy closer to climate change, geoengineering is not unlike coping with higher temperatures and other climate impacts. While no one nowadays would dispute the need to adapt to global warming already baked into the system, it was not too long ago that environmentalists cautioned against even saying "adaptation" out loud. They were worried that doing so would deter from efforts to reduce carbon dioxide in the first place.

Wearing seat belts makes some drivers feel so safe that they drive more recklessly. But that's hardly an argument against seat belt laws. It just means we need to set (and enforce) speed limits, too. In other words: limit carbon emissions.

|||||||||||||||||

If the prospect of injecting millions of tons of tiny, artificially engineered particles into the planet's stratosphere to create a sunshield of sorts doesn't scare you, you haven't been paying attention. Not too surprisingly, it turns out that the vast majority of Americans haven't. In fact, climate polling guru Tony Leiserowitz at Yale has asked Americans, "How much, if anything, have you read or heard about geoengineering as a possible response to climate change." The vast majority, 74 percent, said: "Nothing." Of the other 26 percent who have heard the term, only three percent knew what it meant.

None of that means that we shouldn't take geoengineering seriously. We may be racing past so many climate change tipping points that this kind of planetary "chemotherapy" is already necessary as a Plan B. At the very least, we ought to find out what the full implications are. We can't wait and hope for the best, nor can we hope that the *free-driver* effect won't ever show its full force.

COOLING THE PLANET, FAST AND SLOW

Mount Pinatubo–inspired geoengineering has its appeals, largely because it purports to be fast, cheap, and powerful. But it isn't the only geoengineering option. The basic idea is to reflect more solar radiation back into space. Injecting tiny sulfur-based particles into the stratosphere is just one way, and one of the most daring.

Painting roofs white is sometimes proposed as another. The logic comes down to why winter coats tend to be black, and whites are *en vogue* between Memorial and Labor Days. Black absorbs heat; white radiates it back. This is one reason the melting of Arctic sea ice is so disconcerting. Instead of white surfaces radiating the sun's rays back into space, darker waters and surfaces tend to absorb it, creating vicious circles that heat the planet further. Ubiquitous white roofs in some parts of the Mediterranean contribute to pleasant local microclimates. Some proposals would have us duplicate that effect in urban areas elsewhere. It all sounds pretty good in theory, but there are at least three problems.

For one, we'd need to know the total impact with much more certainty before we go down that path. White roofs reflect more light back into space, but they do so on the surface. The reflected sunlight doesn't escape neatly back into space but now hits soot and all sorts of other air pollutants and particulates, possibly making things worse than before in some polluted cities.

Second is scale. Even if we painted *all* roofs white, globally, we would still only have about a tenth of the impact that one eruption of Mount Pinatubo had by itself.

That brings us to the third fundamental issue: Painting millions of roofs has the exact opposite properties from mimicking Mount Pinatubo. Trying to get millions of people to do something that may benefit the planet comes directly back to the *free-rider* effect. It will be difficult to co-ordinate, unless the act of painting roofs white would pay for itself through, say, decreased need for air-conditioning. It will be all the more difficult if it comes with real costs (and no direct compensation).

There are plenty of options in between Mount Pinatubo–style stratospheric sulfur injections and painting roofs white. An oft-mentioned one is creating artificial clouds or brightening those that already exist. Imagine a fleet of futuristic-looking, satellite-guided ships spraying water into the air to create clouds. It doesn't depend on millions of us doing the right thing. It also doesn't inject anything into the stratosphere that could haunt us once it comes down as pollution. Water vapor is all you'd need, and indeed, some serious proposals have looked into the possibility. In short: it *may* work, emphasis on "may." Brighter clouds *could* lower average temperatures, and the effects may even be targeted regionally.

A regionally targeted intervention could help avoid some of the problems introduced by more global, Mount Pinatubo–style geoengineering. But there could still be plenty of unsavory side effects with enormous implications. The Indian monsoon may be "only" a regional phenomenon, but it's one on which a country of over a billion people depends for its water and food.

As always, it's a matter of trade-offs. Climate change itself will have plenty of unsavory side effects. The question then is not whether geoengineering alone could wreak havoc. (Yes, it could.) The question is whether climate change plus geoengineering is better or worse than unmitigated climate change.

One thing is clear: what you gain in possible precision in any regional geoengineering method, you lose in leverage. Brightening clouds may still be cheaper than avoiding carbon dioxide pollution in the first place, but there are limits on what it might accomplish. Mount Pinatubo–style geoengineering has much greater leverage and, thus, overall impact—for better or for worse.

||||||||||||||||||

All these geoengineering methods—from Mount Pinatubo to brightening clouds to painting roofs white—have one thing in common: they don't touch the carbon dioxide already up in the air. That makes them potentially cheap. It also means that they avoid tackling the root of the problem.

Cue "carbon dioxide removal" (CDR), confusingly also called "direct carbon removal" (DCR). It, in turn, comes under various guises. "Air capture" takes carbon dioxide out of the air and, for example, buries it underground. "Carbon capture and storage" stops carbon dioxide from entering the air in the first place by scrubbing it out of smokestacks and treating it in a way so it doesn't escape back into the air. "Ocean fertilization" does just what the name suggests: dump iron or other nutrients into surface waters to make them more fertile grounds for natural carbon dioxide uptake. "Biochar" is a fancy term for charcoal and may have effects similar to other approaches: take carbon dioxide out of the air and prevent it from escaping back. You could even put plain old trees into that category; trees take carbon out of the atmosphere naturally as they grow. In fact, there's often little that humans need to do other than get out of the way. Nature takes care of reforestation in many situations, as long as there's no added interference.

Opinions differ on the effectiveness of any of these methods. Opinions also differ on whether they should even be called "geoengineering." They are methods of geoengineering, in the sense that someone would intentionally try to engineer the earth's atmosphere on a grand scale. It's precisely the issue of scale, though, that's up for debate.

Most of these approaches run head-on into the free-rider problem. It either takes coordinating the actions of millions to have an impact, or it takes one person to spend so much money that he's unlikely to do so. In other words, few of these approaches share the properties that make Mount Pinatubo–style geoengineering so unique. They have a lot less leverage. They are often expensive. They are slow. In fact, they look much more like reducing carbon emissions in the first place than geoengineering.

Of course, we aren't saying that the world shouldn't consider any of these approaches. We should be growing more trees, say, almost regardless of their climate impact.

The same may go for painting roofs white to lower air-conditioning costs. But that doesn't mean we should lump these methodologies together with Mount Pinatubo–style geoengineering. That goes both for roof whitening and for any form of "carbon dioxide removal"—from trees to ocean fertilization to carbon capture on smokestacks. All are important. None is in the same category as shooting sulfur or other tiny particles into the stratosphere directly.

ADDICTED TO SPEED

Everyone's very first cup of coffee tastes bitter, no matter how much sugar and milk you add. The second cup in your life may already be a bit more pleasurable. By the twentieth, you may think you are still not addicted and that you can easily skip twenty-one and twenty-two. The twenty-third cup is when you discover cappuccino. And whatever you do, by the hundredth cup in your life, you are hooked. Stopping is no longer an option.

Mimicking Mount Pinatubo to cool down the planet follows a similar pattern. The first attempts at deploying geoengineering may well fail. By the twentieth, we may be ready to take a break. By the twenty-third we'll discover a more refined technology, and sooner or later it's impossible to stop.

Start-up woes come with the territory. It's the addiction component that's a worrisome additional aspect of Mount Pinatubo–style geoengineering.

In 1991, Mount Pinatubo cancelled out 0.5°C (0.9°F) of warming. When most of the remaining sulfur dioxide from Mount Pinatubo washed out of the atmosphere and the cooling effects of the volcano wore off two years later, temperatures jumped back by those same 0.5°C (0.9°F) and resumed growing thereafter.

By now, temperatures have risen by 0.8°C (1.4°F) since preindustrial times. If we wanted to erase that difference using geoengineering and then suddenly had to stop, temperatures would jump back by 0.8°C (1.4°F). By 2100, this jump back could be on the order of up to 3–5°C (5.4–9°F), if we haven't severely restricted emissions long before then. Scientists don't know what would happen with a jump of 0.8°C (1.4°F). They are pretty sure jumping 3–5°C (5.4–9°F) would create serious problems. Slow warming of 3–5°C (5.4–9°F) by the next century or so would be bad enough. A sudden jump from abruptly ending geoengineering would result in all sorts of additional issues. Moving major agricultural areas from Kansas to Canada is disruptive in every sense of the word. But doing it over a century is at least possible. Having to do it within a year or a decade is hard to imagine. Or at least it will cost exponentially more.

In that case, costs in dollars and cents may be the least of our worries. A bigger fear may well be that stopping a continuous Mount Pinatubo–style intervention won't come in isolation: Whatever geoengineering we could ever do on a sustained, global scale would require almost unprecedented global governance systems. It's easy to imagine that falling apart for all sorts of reasons.

War is one. Even your standard political fickleness would do. A regime change anywhere could jeopardize the global agreement for everyone, which may well lead right back to war. Given that militaries everywhere already consider global warming itself a national security threat, the world had better prepare for all eventualities. The addiction component of Mount Pinatubo–style geoengineering and its vulnerability to interruption may turn out to be its biggest problem yet.

WALK BEFORE YOU RUN,
RESEARCH BEFORE YOU DEPLOY

Fortunately, we aren't close yet to anyone seriously proposing to *deploy* geoengineering at scale today. Even David
Keith, the author of *A Case for Climate Engineering*, says that
he wouldn't vote for geoengineering deployment now. We
are, however, way past the time when serious people are
proposing to *research* geoengineering, Keith among them.

Asilomar 2.0 was chock full of researchers and engineers
who are actively looking into the *how*, hence their desire
for guidelines on just how to move forward with their research. Plenty of options are already on the lab table. Researchers want to know how far they can go in testing and
refining their methods in the real world.

One real hurdle of researching with the entire planet
as your test subject is discerning when a signal rises above
the noise. The bigger the experiment, the easier it is to
detect the effects. But the lines between research and deployment quickly get blurry. Even studying the full effects
of Mount Pinatubo has proven difficult for precisely this
signal-versus-noise issue. Putting 20 million tons of sulfur
dioxide into the atmosphere constitutes a major disrup-
tion. It's clear that little else could have contributed to the
global cooling effect of 0.5°C (0.9°F) in the subsequent
year. Similarly, reasonable atmospheric mechanisms could
explain how adding carbon dioxide and then dimming the
lights a bit by means of geoengineering means less rainfall around the globe. That alone would explain a higher
likelihood for droughts. But despite the general advances
of attribution science, linking any one particular flood or
drought to single geoengineering interventions is fraught
with difficulties.

CAUSALITIES, SHMAUSALITIES

Public opinion does not react well to mistakes and unintended consequences. And geoengineering is nothing if not fraught with the potential for errors. But not all errors are created equal.

There's a big difference between errors of omission and commission: Driving by the scene of a car crash is bad, but it's not as bad as causing the crash in the first place.

Driving by is the omission. It may be illegal for anyone with an "M.D." printed on their license plate, which comes with some privileges but also increased responsibilities. But even doctors swear only to "do no harm." They don't give an oath to save every human being everywhere.

Commission is worse. Causing the crash is bad no matter how you look at it.

It's one thing to study the effects of Mount Pinatubo. The harm has already been done. No one could have prevented the eruption. And it has turned out to be the best-studied major volcanic eruption ever. Let's use that for all it's worth. (Not studying it to its fullest may be an error of omission all by itself.)

It's similarly easy to model Mount Pinatubo–style interventions on a computer. It's cheap. It's low-impact. It may divert attention away from other pursuits aimed at limiting carbon dioxide emissions, but that's about the worst that can happen. Little harm is done by a graduate student spending extra time on a Saturday in the lab running one more simulation.

It would be very different for scientists to go out and *intentionally* experiment with the atmosphere. Now we are in the world of *commission*, and a complicated one at that.

It may not be sensible to link a failed harvest to a small experiment halfway around the world that barely produces enough data to identify the signal from all other climatic noise, but that won't really matter. The burden of proof in the court of public opinion will be on those running the experiment.

|||||||||||||||||

Let's just take a quick step back to try to put it all into perspective. The greenhouse effect has been a fact of science since the 1800s. The term "global warming" has been around since 1975. The basic science has been settled for decades. We have no excuse to think that using our atmosphere as a sewer for our carbon emissions isn't uneconomic, unethical, or worse. All seven billion of us—especially the one billion high emitters—are committing sins of commission every single day. The effects of our collective actions may end up being catastrophic and may end up killing people. No single person is guilty of any single climate change–related death, but collectively we all are.

Now contrast that with a group of scientists committed to finding a way out of the global warming mess. They understand the science. They understand that the *free-rider* effect discourages society from acting on time to curtail emissions. They understand that the siren call of the *free-driver* effect is pushing us toward the all-too-alluring quick fix. They are working on trying to understand if and how that fix could indeed work, and how it could be made safe for the planet to even consider using.

We are not trying to excuse any and all scientific (mis)conduct. Like all other walks of life, science has plenty of misfits, mercenaries, and ill-intentioned missionaries. Not all budding geoengineers should be considered heroes, but at the very least they shouldn't all be considered James Bond–style villains until proven otherwise. Scientists themselves are asking for guidance, as the Asilomar 2.0 meeting and many other similar efforts make clear. They know they can't go this one alone, even if they wanted to. And most don't want to.

AN ALMOST PRACTICAL PROPOSAL

One of the more sensible proposals of what to do next comes straight from David Keith. It starts with the "m" word: moratorium.

There's a lot to that position. Scientists themselves need to realize that there's a clear danger for the science to run ahead of the public conversation. The only way to stop that is a self-imposed moratorium. In "End the Deadlock on Governance of Geoengineering Research," Keith, together with Ted Parson, proposes to guide research on geoengineering by following three simple steps:

> First, accept that there must be limits.
> Second, flat out declare a moratorium on all research above a certain size.
> Third, set a clear and very small threshold below which research may proceed.

In a sense, these three steps just formalize the natural progression of research: start small; experiment; evaluate; tackle the next challenge. By declaring such a public "moratorium," their thinking goes, at least the smaller experiments would be acceptable. Of course, everything depends on where that line is drawn. Parson and Keith have not specified where their "clear and very small threshold" is. It must be very small indeed: zero is a good starting point.

In all of this, we also need to realize that humans are already spewing massive amounts of pollution into the atmosphere, including the very substances that some geo-engineers propose to use to help cool the planet. Research that has a fraction of the impact of any one jet engine is one thing. Research large enough to have detectable impact beyond the narrow confines of the experiment is a clear nonstarter. In any case, the goal must be a much better understanding of the full set of benefits and costs—and especially the costs of geoengineering.

The fact that Mount Pinatubo–style geoengineering represents a *free-driver* problem means that sooner or later it will be hard to maintain any such self-imposed moratorium. As long as there are only a dozen or so geoengineers on the planet, all of them know and respect each other, and all of them agree on the importance of not letting the science get ahead of the public, fine. But it's not hard to imagine some scientist somewhere wanting to leave a mark and go it alone.

There's a larger question at work here, too. Moratorium to what end? Eventually, we may need to have a conversation about lifting the moratorium. What comes then? How do we decide to lift the moratorium? Who will decide?

CHAPTER 6

007

CUE TITLE MUSIC. *Sean Connery look-alike enters bar, downs martini—shaken, not stirred. He stands calmly as the bar around him erupts in chaos following an explosion that may or may not have been caused by Mr. Bond himself.*

MAN AT BAR: There's a plane leaving for Miami in an hour.
BOND: I'll be on it. But first I have some unfinished business to attend to.

||||||||||||||||||

Fast forward through bedroom scene. Cut to ritzy Miami Beach resort. Palm trees. Infinity swimming pool overlooking ocean. The infinity pool used to tower over the ocean at high tide. By now, near-weekly storms whip seawater into the pool.

OWNER: You won't believe how much these storms are costing us. And we're shut for two weeks every year. In high season. It can't go on like this.
POLITICO: Tell me about it. The entire district is hurting. Three streets of houses gone last year. The richest donors all left. Moved uphill to Miller's. He's getting all the love. I'm toast. I'm pleading with you not to move. Pleading.
OWNER: I'm not going anywhere. Can't stand Miller. Don't worry.

Long pause.

Owner: There's just one thing …

IIIIIIIIIIIIIIIII

UK Secret Service Headquarters. Bond flipping through charts with his boss M.

    M: That's the problem. It could do a lot of good. Imagine wiping out two centuries of global warming, dialing temperatures back to before we started burning coal *en masse*. But in the wrong hands, it's a weapon.

    Bond: Cost?

    M: Peanuts, at least for this guy.

    Bond: But why do it?

    M: Money. It's always money. He's been buying up beach-front resorts for scrap value …

    Bond: … while everyone else has been fleeing inland. Bastard. Smart bastard.

IIIIIIIIIIIIIIIII

Cut to aerial shot of wood paneled room on 122nd floor of United Nations headquarters in Abu Dhabi, United Arab Emirates, fortified by the world's strongest seawall to hold back the rising sea. Leaders of the twenty most powerful nations hash out a response to Indonesia, discovered last year to have been experimenting with what's since been dubbed "Pinatubo Two": mimicking the global cooling effect of Mount Pinatubo by intentionally shooting sulfur into the stratosphere.

    U.S. Secretary of State: Unacceptable. There's no legal basis for any of this.

    Indonesia: We've had a state of emergency for over a decade now. Land disappearing. Crop failures. We had 30,000 deaths last year alone, and two million people on the move. All due to bigger storms and rising seas.

INDIA: Climate refugees.

U.S.: Beg your pardon?

INDONESIA: Refugees. Climate refugees. Over a hundred abandoned islands, tens of thousands of refugees.

U.S.: OK. So, where are we?

INDONESIA: We've concluded phase two of our three-phase research program—flying progressively more planes with more sulfuric acid vapor into the stratosphere. It's all been done to the highest international standards. Graduate students at your very own Harvard Jakarta campus are helping us analyze the data. The funding comes from our National Science Foundation. Some of the best experts globally are serving as external advisors. We were just about to launch phase three, full deployment . . .

U.S.: . . . but a security breach revealed blueprints of your efforts. Someone out there stole them and is going rogue, quadrupling the dose of sulfuric acid. And it's all been undetected for almost a year. Yes, yes. What's new?

INDONESIA: First data are in.

Indonesian representative points to a chart, which eerily looks like a "hockey stick" lying there with its long handle on the ground and its blade sticking up: small amounts of tiny sulfur particles in the stratosphere cool global temperature by a bit; higher doses get a significantly larger response.

INDONESIA: There's something we just don't understand. There's a break point after tripling the dosage. It's just off the charts.

U.S.: How sure are you?

INDONESIA: Sure enough to have convened all of you here today. We've since pulled back completely: no more sulfur going into the stratosphere. But someone has picked up our slack, and then some.

Zoom out. Discussion continues.

||||||||||||||||

Meanwhile, at UK Secret Service Headquarters, Bond and M are looking at video recording of the discussion.

BOND: And now we are talking ten times?

M: Ten times the original dose of sulfur injected into the stratosphere.

BOND: And we can't tell who because too many private plane operators globally could be doing it all by themselves.

M: Yes. Except . . .

BOND: Except?

M: Except we have one lead.

||||||||||||||||

Bond's jet lands in Rio de Janeiro, Brazil. Checks into hotel, picks up phone when entering room.

BOND: Excellent. Twenty-two hundred. Upstairs bar.

||||||||||||||||

Upstairs bar. Wall clock flips to "22:00." Bond enters.

BOND: Two jets. Well done.

CARIOCA: Private jets.

BOND: Plane condition?

CARIOCA: Perfect. Latest model. Uh, fully functioning.

BOND: But?

CARIOCA: No seats. No furnishings. Nothing. Just . . .

BOND: . . . a trap door.

CARIOCA: Amazing how much trouble criminals go to these days to dispose of bodies.

Bond looks on, finishes drink. Zoom to clock: 22:02.

BOND: Pleasure.

||||||||||||||||||

Aerial shot of Miami Beach resort from opening scene. Owner talking to group of staff, gets interrupted by phone call.

OWNER: How many? Two? Hah, two of a hundred. Call me when there's news.

||||||||||||||||||

We aren't the first to call attention to the possibility of a "Greenfinger" taking Mount Pinatubo–style geoengineering into his own hands. Political scientist David Victor coined the term to describe just this possibility. It may seem as unlikely as any Ian Fleming novel about the world's top fictional spy, but it's not entirely far-fetched.

Spoiler alert: The fictitious, rich hotel owner turns out to be behind a global scheme to rig the well-intentioned, expert-controlled Indonesian-led geoengineering trial. It may be impossible to imagine how anyone could steal that much sulfur undetected, or whether the tiny particles of choice at that time in the future would even be sulfur-based. Leave that to the screen writers.

There are also plenty of political and other question marks. Would the modified twenty-member UN Security Council have stepped in before Indonesia was even able to conduct its Pinatubo Two experiment for almost a decade? Could it have been done undetected? Would it have been openly decried but privately tolerated and even welcomed?

A few points are clear. Depositing millions of tons of sulfate particles into the stratosphere with high-flying jets is well within the reach of one country, especially one the size of Indonesia. The motivation would be equally clear. Bangladesh is often used in this instance: a low-lying country disappearing under the rising seas. Tens of millions of people on the move. Tens of millions more depend on rivers flowing in East Asia. Millions depend on various other climatic patterns that have been relatively stable for millennia and have enabled civilization as we know it today to exist. Disrupt those patterns, and it may well trigger the desire to intervene. Bangladesh's or Indonesia's or India's or China's national security advisors would be remiss not to consider the possibility.

We don't need to point fingers to an individual country; any large country—developing or not—would have the capabilities. It may be nearly impossible to get a specific geoengineering method past a U.S. Environmental Protection Agency review. It may even be impossible to do so in a democracy like India or Indonesia. Or it may not. The point is that it's not inconceivable. The *free-driver* effect all but guarantees that it will happen one day.

CLIMATE WARS

We can play all kinds of theoretical games of political back-and-forth to see where the pieces might land. Imagine that climate change disrupts the Indian monsoon and, thus, the food source of tens of millions on the subcontinent. Geoengineering, in turn, could disrupt East Asian rivers, and, thus, the food source of tens of millions of Chinese. What if optimizing geoengineering for India hurts China, and vice versa? Would we want a geoengineering match between two nuclear powers, each with over a billion people?

What if there were Mount Pinatubo–style technologies to cool the planet, and equally effective antidotes to warm it? In fact, these fast-acting warming technologies already exist as well. Some industrially produced gases like hydrofluorocarbons (HFCs) have a hundred to over 10,000 times more warming potential than carbon dioxide in the short run. Let the climate war games begin.

Imagine a scenario where one country threatens to counter any unilateral geoengineering attempts. Acting on that threat would likely mean a worse outcome for everyone involved: geoengineering plus counter-geoengineering may balance global average temperatures, however imperfectly. Yet both would likely come with their own sets of unpleasant side effects, which are unlikely to cancel each other out. If anything, tiny sulfur-based particles could interact with HFCs in entirely unanticipated ways.

It's also possible to imagine nonlinear responses. Ten times the dose, 1,000 times the response—the example used in the Greenfinger plot—may be farfetched, but not entirely so. We don't know the leverage of geoengineering at these extreme levels. But it's not hard to imagine a scenario where a large decrease in total solar radiation could result in temperature decreases to below preindustrial levels. Runaway global warming is bad. Creating an artificial ice age would not be great, either.

SULFUR DOESN'T HAVE SEX

James Bond–style fiction or not, one thing is clear: geoengineering still takes human intervention. Sulfate particles don't have sex. They don't reproduce by themselves, and so they can't produce some kind of runaway geoengineering scenario if left unattended. It's humans that cause the havoc, not nature. If one stops pumping tiny particles into the stratosphere, what's there will wash out after a few months, and Mount Pinatubo–style geoengineering will have stopped.

Stopping, by itself, will come with costs. The addiction analogies for geoengineering are apt. But the fears on display at the self-described "Asilomar 2.0" meetings on geoengineering in 2010 are in a different category from those at the original Asilomar meetings in 1975 around biotechnology. There was then and still is today the remote fear that recombinant DNA research somehow might result in reproducing organisms that wreak havoc all by themselves. We discuss and ultimately dismiss some of these biotechnology concerns in chapter 4, invoking the specter of natural selection: it's unlikely for scientists to one-up nature, which has tried untold DNA combinations herself. But in biotechnology the theoretical possibility at least exists. That possibility does not exist for geoengineering. Scientists and climate engineers have lots of reasons to worry; particles reproducing isn't one of them.

WHAT IF GEOENGINEERING DOES WORK?

You can paint any number of scary scenarios and what-ifs where things can go horribly wrong. Geoengineering may be bad. Really bad. But of course, we also know that traditional pollution is bad. Perhaps even worse. Sulfate particles injected into the stratosphere will eventually wash out, notably with bad health effects, killing thousands of people globally. Traditional outdoor air pollution, meanwhile, is killing over 3.5 million people globally this year.

What if geoengineering does, in fact, work to reduce some of the worst global warming effects that burning coal has wrought and has yet to bring? For the time being, the best response is still to scream even louder for decreasing carbon dioxide pollution in the first place. Full stop. But looking many years and decades out, other options may have to be considered.

IT'S A RISK-RISK WORLD

At the heart of the geoengineering discussion is the distinction between errors of omission and errors of commission: driving by the car crash versus causing one.

Omitting sensible climate policy may be less bad than committing errors in designing it. Avoiding blame ranks highly in the minds of politicians.

Omitting geoengineering then may be less bad than committing errors in designing it. That may, in part, explain why we see too little geoengineering research.

The line isn't always clearly drawn. Ultimately, what is omission and what is commission depends on your point of view. Is the world now committing errors of omission by not having sensible climate policies, or are we committing errors of commission by polluting too much? On top of that, there's surely a sliding scale. If racing to defuse a time bomb could prevent 1,000 people from dying but may cause one death along the way, it's unclear whether ignoring the ticking bomb (error of omission) will indeed be less bad than causing that one death (error of commission). What about a ratio of 1,000,000 to 1? Or 1,000,000,000 to 1?

There must be a point somewhere between that 1,000,000,000-to-1 and a 1-to-1 ratio where errors of omission become as bad as errors of commission. If geo-engineering truly has the potential to save or improve millions of lives, at some point it may be worth the trade-off.

But who decides when the trade-off becomes worth it? It can't just be a handful of scientists. We also can't wait for all 190-odd nations to agree on a course of action. Some-one will surely take matters into their own hands before the world reaches that point of agreement.

||||||||||||||||||

We cannot tell you how much geoengineering should be done. But a bit more logic could at least bring us closer to a decision criterion—a voting rule for answering that question.

If we put geoengineering up for a vote at the U.N. Frame-work Convention on Climate Change, we'd need unanim-ity. Any single country can block progress in that body. No surprise then that progress is glacial. By contrast, the U.S. House of Representative needs a simple majority. For all practical purposes, the U.S. Senate requires a 60:40 major-ity to overcome filibusters and get anything done. For a treaty, it's 67:33. We could have endless debates around which voting rule ought to prevail for the world to decide on the optimal level of climate intervention.

Here's another proposal: zero in on the comparison of errors of commission to errors of omission. If you think an error of commission is twice as bad as an error of omission—or twice as likely—the ideal voting rule re-quires a two-thirds majority to proceed: $2/(2+1)$. If you think it's three times as bad, it's three-quarters: $3/(3+1)$.

If you think it's four times as bad, the voting rule ought to require a four-fifth majority: $4/(4+1)$. You get the pattern.

The mathematical derivation behind that formula may be too complex for words, but the logic is quite simple: If errors of commission are no worse than errors of omission, go do geoengineering. If errors of commission count for a lot, don't. More specifically, require a larger majority to agree before going ahead with geoengineering.

There are plenty of ways to criticize this simple formula. First and foremost, could it be just a tad bit too rational? It does assume that society is interested in doing the greatest good for the greatest number of people. It may be hard to quibble with that proposition in theory, but the practice often looks rather different. Still, the basic underlying logic should hold in this often illogical world of ours: the more we are worried about the negative side effects of geoengineering relative to not doing enough about climate change in the first place, the more people should have to agree to start any kind of geoengineering intervention. That's not a revolutionary statement. The formula just brings it all the more into focus.

Together with an (initial) moratorium on all research above a certain scale, this formula could then serve as a starting point for our governance conversations. Since we know that the trade-off between errors of commission (geoengineering gone awry) and errors of omission (runaway global warming without geoengineering to address the worst of it) is at the heart of the matter, let's focus on the trade-off of these errors directly.

Deciding on a voting rule for actual geoengineering, of course, is jumping the gun quite a bit. First we need to focus on research. There, too, looking at potential errors could provide some guidance. It's what happens when things go wrong that may define the outcome. What's the chance of 10, 100, or 1,000 or more people dying as a direct result of a particular intervention? What—if any—existential risks are we introducing through geoengineering itself?

FULL BENEFITS VERSUS COSTS, WITH SOME UNKNOWNS AND UNKNOWABLES MIXED IN

Here's perhaps the most important question of them all: what are the true social costs of geoengineering—with all the potential nasty side effects of shooting sulfate particles into the stratosphere—and how do they compare to the potential benefits?

The $1 to $10 billion estimate of the narrow engineering cost alone leads us to the *free-driver* effect, but that doesn't say anything about the more inclusive true social costs. For all we know right now, the indirect costs of potential side effects could end up dwarfing any of the purported benefits. That, in some sense, is the worst of all possible worlds: Believing that geoengineering will be cheap and could do wonders, if in reality it isn't and won't.

It's also where we are coming full circle: Ignoring full costs and consequences of our actions has been causing the climate problem in the first place. Let's look at the full costs of any and all proposed solutions, and focus in particular on the unknowns and unknowables.

CHAPTER 7

What You Can Do

Y OUR VOTE DOESN'T COUNT.

Put a group of economists in a room to debate the wisdom and virtue of individual action, and soon they'll be debating the value of casting one's vote: zero—in some strict, narrow, "economic" sense of the world.

It's a tough pill to swallow and runs counter to every call for civic duty, but we don't say this lightly. The chance of your one vote making the ultimate difference is so small we might as well call it zero. Some of the best research on this topic—by a team including Nate Silver of baseball statistics and, more recently, of FiveThirtyEight electoral prediction fame—put the probability of your vote making a difference in a U.S. presidential election at 1 in 60 million. And that figure includes the 2000 George W. Bush–Al Gore match-up in Florida. These are steep odds, to put it mildly. Even if your candidate were able to boost U.S. gross domestic product by ¼ percent in a given year, and we assume a very close election, your personal benefit of casting the decisive vote would still only be a tiny fraction of a penny. In other words: zero.

We can't leave things there. It would be both rather depressing and also rather narrow-minded. Maybe statistics and economics alone aren't the right tools with which to analyze one's personal action. Ethics, for one, plays an important role.

WHY VOTE

Self-declared "rational" economists may continue to shake their heads in private and joke about how voting is one of these unexplained mysteries. It's not a mystery for the rest of us. We all know that voting is the right thing to do. Our military men and women pay with their lives for us to be able to cast that vote. It's a sacred right. It's the epitome of democracy. Not voting shows contempt for American—for human—values. We shouldn't just vote. We should *rock the vote*. Or at least we should display a prominent sticker declaring that we have voted, thereby coaxing others into doing so.

Your personal monetary gain may be zero, but that's beside the point. The point is to do the right thing, and voting is as simple as that gets. It doesn't require you to take your family to a soup kitchen on Christmas morning to volunteer. You don't need to pay extra to do it (ever since poll taxes were outlawed). Some employers even give you the day off. And you can express your opinion without having to do so publicly. You don't have to tell anyone whom you voted for, as long as you do vote. Civic duty fulfilled.

Academics have a way of complicating things quite a bit. Here's a shortened version of what Jason Brennan describes as "the folk theory of voting ethics":

1. Each citizen has a civic duty to vote.
2. Any good faith vote is morally acceptable. At the very least, it is better to vote than to abstain.
3. It is inherently wrong to buy or sell one's vote.

Brennan then spends 200 pages destroying that folk theorem and arrives at a more complex ethical justification for

voting. He could even live with people buying and selling their vote, but any old vote won't do. If you don't vote
for the common good over your own narrow self-interest,
don't vote at all.

In other words: Your civic duty isn't just to vote, it's to
vote well. It's a tough argument with which to disagree.
Vote for a cause larger than yourself. Vote for those who
promise more than just to further their own agenda (or
yours!). Vote for those who seek to look out for society at
large.

Whatever that may mean in a specific case, it clearly goes
beyond the not-sure-I-should-vote-I'd-rather-watch-TV reasoning. Get up and vote; it's the right thing to do. And
don't just vote for the sake of voting. Vote as an informed
citizen. *Vote well*.

That means thinking through the questions we're asking
in this book, and then seriously asking yourself whether to
vote for candidates who will act on climate change.

WHY RECYCLE, BIKE, AND EAT LESS MEAT

Shift gears to reducing, reusing, and recycling, the man-
tra of every good environmentalist. The thinking there is
roughly the same as for voting. Your single act of kindness
isn't going to change the course of history. Recycling won't
stop global warming. One of us wrote an entire book ti-
tled: *But Will the Planet Notice?*

No, it won't.

The math is about as clear as can be, all without having to go through Nate Silver–type reasoning to guess how
likely it is that your vote could decide a U.S. presidential
election. Reducing your own carbon footprint to zero is a

noble gesture, but it's less than a drop in the bucket. Quite literally: the standard U.S. bucket holds about 300,000 drops; but you are one in over 300,000,000 as an American, and you are one in seven billion as a human being.

Every little bit doesn't always help. In the words of David MacKay analyzing the implications for the wider energy system, "if everyone does a little, *we'll achieve only a little.*" So why go green at all? Because it's the right thing to do. It's also how we learn the values that we have to apply on a much larger scale to tackle climate change.

Recycle. Bike to work. Eat less meat. Maybe go all the way and turn vegetarian. Teach your kids to do the same, and to turn off the water while brushing their teeth. It's good for you. It's good for those around you. It's the right thing to do.

But do it well. Don't just vote. *Vote well.* Don't just recycle. *Recycle well.*

RECYCLE *WELL*

If individual, inherently moral acts of environmental stewardship—like recycling—lead to better policies, sign us up. The goal, in the end, is to enact the best overall policies that will guide market forces in the right direction. So if asking one more person to recycle more is the foot in the door for their going to the polls and voting for the right policies that are in the common interest, great. Ask people to "go green" in some small way like bringing a canvas bag to the store, and they may feel a greater moral obligation to do something about larger environmental issues. Psychologists call it "self-perception theory": see yourself as greener, vote greener.

Cue the virtuous circle of civic engagement, informed behavioral change, and all-around good things for a better planet: Voting well leads to better policies, which allow for a more enlightened citizenry; a more enlightened citizenry, in turn, leads to more people voting well. Recycling well leads to better environmental policies, which allow for a more environmentally enlightened citizenry; a more environmentally enlightened citizenry, in turn, leads to more people recycling well.

Call it the *Copenhagen Theory of Change*. Danes didn't wake up one day and decide to bike to work en masse through the bitter cold of northern Europe. Nor did Copenhagen's Lord Mayor wake up one day and decide to install a sufficient number of bike paths to get his residents out of their cars and onto bikes. Cars had dominated Copenhagen much like most other European cities for decades. It took the fuel crisis of the 1970s, increased environmentalism, and years of activism to go from "car-free Sundays" to over 50 percent of Copenhageners commuting by bike every day.

And biking isn't unique. The Voting Rights Act wasn't passed overnight. It took years of all sorts of action—from the early sit-ins to the Selma marches. The U.S. environmental movement, which sparked the "environmental decade" of the 1970s, followed a similar path. Years of self-reinforcing activism eventually led to the necessary legislative changes, and the debate didn't end there.

Time is the all-important factor. We once had decades to turn the climate ship around. Not anymore. That makes it all the more important to get our theory of change right. It's also where we return to our recurring theme of trade-offs, this time as it pertains to *recycling well*.

Economists deem the existence of trade-offs to be self-evident common sense. Psychologists add another twist to it, turning the effects of "see yourself as greener, vote greener" on its head. Call it the "crowding-out bias." The threat of climate change motivates people to act—but only up to a point. In the extreme version of this effect, the "single-action bias," people may do only one thing, like recycle, or put a solar panel on their roof, or buy a "green" product. This doesn't necessarily mean that anyone, in fact, believes that one step is indeed enough to stop climate change, but that one step may be enough to assuage their worries and lead them to move on. Yes, the climate is changing, but women are still dying in childbirth. There are other problems to worry about, too, and now I've done my part for climate change.

Economists are instinctively more comfortable with this *crowding-out bias* view of the world than the one supporting the self-perception theory, a.k.a. the *Copenhagen Theory of Change*. After all, trade-offs often lead people to substitute one action for another. That's particularly troubling when people substitute single, individual actions—like recycling—for larger policy actions—like voting. This phenomenon has been surprisingly poorly studied so far.

We know quite a bit about the mechanisms from collective action back down to individual ones. Setting the right incentives—paying people to do certain things—sometimes crowds out virtuous behavior. Pay people to donate blood, and watch blood donations go down, at least among women. Men seem to have no qualms about being paid for their donations, and women, too, increase their donations once again when the money is given to charity rather than paid out to them.

We also know a bit about substitution among individual actions. Ask people to voluntarily pay more for "green electricity" and watch some increase their electricity consumption as a result.

Both of these mechanisms support the *crowding-out bias* view of the world, where one green deed doesn't necessarily beget another but may indeed be a hindrance in light of trade-offs in people's everyday behavior. However, we know little about whether the crowding-out bias does, in fact, extend from individual to collective action.

No one wants the *crowding-out bias* to dominate. It's something to avoid and overcome. If you catch yourself recycling that paper cup and thinking you've solved global warming for the day, think again. If you catch yourself buying those voluntary carbon offsets for the cross-country flight, feeling better about flying, and as a result doing more of it, that's not quite in the spirit of the exercise either. "The hotel changes my towels only when I throw them on the floor *and* the airline lets me spend $20 extra to offset my carbon pollution? Eco-vacation, here I come!"

None of that is all that far-fetched, even for the most committed of environmentalists. You can't do it all. Plenty of environmentalists who recycle, refuse meat, don't drive, and generally try to do it all the green way still commit various other, often more significant carbon sins. Flying is a prime example.

THE SKY IS THE LIMIT

Even the most committed canvas-bag-carrying, reusable-water-bottle-sipping environmentalists often draw a line with flying. You can and should take the train from New York to Washington, D.C., but Miami to Seattle is another matter, and Atlanta to Beijing is impossible. Such is the lot of the transcontinental environmentalist: talks to give; meetings to hold; melting glaciers to see firsthand. Video conferences may work in lieu of boarding yet another flight, but sometimes you simply can't phone it in. Diplomacy, as we all know, happens over after-dinner drinks.

Not going on that business trip won't do. It's the classic example of the not-so-invisible hand of the market at work. If you generously volunteer not to board the flight in order to cut down your carbon footprint, the planet won't notice your sacrifice. Your competitor will.

You can always pay someone to plant a tree or capture the methane from a manure pit on your behalf to offset the pollution your flight causes, and you should. We clearly should be planting more trees and covering more manure pits. But none of that amounts to the type of change that's truly needed. That kind of change can happen only on the policy level.

We can look to the European Union for some of the answers. Its emissions trading system has been covering domestic flights since January 2012. Passengers on the typical flight within the EU pay for part of the carbon pollution they cause. Current prices average out to about $2 per ton of carbon dioxide. That's not nearly enough to cover the true pollution costs of each flight, compared to the $40 or more necessary to capture the full costs. But it's a significant start and a crucial step up from voluntary action. Passengers who pay these small amounts can now take their trips with an ever-so-slightly clearer conscience. Instead of reaping all the benefits of showing up for that client meeting while seven billion others pay for the cost of pollution, everyone will begin to pay for their own pollution costs and will thus be moved to change their actions. The goal, of course, should be to broaden and deepen the system: incorporate the full pollution costs, and not just for intra-European flights.

That's where the International Civil Aviation Organization ought to come in and get serious about instituting a truly global approach to aviation emissions. The level of ambition for such a global approach is all-important, though the principle is clear. Sir Richard Branson got it exactly right: "I think a global carbon tax is screaming—blindingly obvious and should have been introduced 15 years ago, and that would have been completely fair. Every single airline in the world would have been treated in the same way, every single shipping company. . . . [A]s an airline owner, I'm sure I'll get told off when I get home . . . but there should be a fair global tax with everybody taking a little bit of pain. It's not massive. And if that happened, we would get on top of the problem."

Amen.

Climate policy isn't rocket science. It's harder. But the solution is Bransonianly obvious: price carbon. The question is how to get there.

If "self-perception theory"—the *Copenhagen Theory of Change*—wins the day, every little bit builds toward the next and eventually leads to half of Copenhageners biking to work *and* strong national policies to guide the masses toward a low-carbon, high-efficiency world. In that case, recycling, reusing, and buying voluntary carbon offsets may well lead to real change, quickly.

If *crowding-out bias* wins the day, many too-direct appeals to our better nature may be counterproductive. That may be particularly true for those in the middle of the political spectrum who will ultimately be deciding overall policy. It's easy to convince environmentalists to recycle more, but they will already be voting for strong climate policy no matter what. It's the ones in the middle who need to be convinced.

It's clear that neither theory of change will apply in all situations. The world is much more complex than either simple mechanism would suggest. One thing is clear: Fight *crowding-out bias* at all cost. And if you have to make a choice between recycling and voting for a price on carbon, choose voting.

STEP #1: SCREAM

What are the things you *can* do? For one, don't trust any list that gives you ten things you can do to stop global warming. You can't stop it yourself. If biking to work and turning down the air conditioning might inspire your friends and coworkers and so help build a movement, great. But those actions alone won't heal the atmosphere. Recall the drop-in-the-bucket analogy. There's a fine line between simplistic single actions and doing what counts. You can't stop global warming, but what if a big box retailer announces it will green its supply chain to remove 20 million tons of carbon dioxide pollution by 2015? What if a major airline used voluntary carbon offsets not just as a marketing tool but to evoke real change, perhaps even because it would benefit directly from a global carbon pricing system as its fleet is younger and more efficient, and hence pollutes less than the competition? What if the decision is about building a pipeline to transport particularly dirty tar sands oil from Canada to refineries in the Gulf of Mexico?

The simple answer is that laggards can't be choosers. If greening your supply chain isn't just good for the planet but also good for business, green it. It's a win-win. The same goes for building a new pipeline—or rather not building one. If an honest benefit-cost calculation shows it's not worth the price the planet will have to pay, the decision is clear. But once again, it's the next step that likely matters more: If the first step leads to more momentum and more action down the line, take it. If it stops us from doing more, stop and think. Trade-offs matter. They matter for individual action. They also matter for policy.

In the end, a democratically elected government does —up to a point, over time—do as its citizens want. That's where activists enter. If getting arrested in front of the White House makes it clearer to the president that we want action, it may well prompt that action. Civil Rights had Malcolm X *and* Martin Luther King *and* Rosa Parks, each pursuing their own strategy. Some may have decided against one or another action because they figured it could backfire or not amount to enough. But eventually all could take credit when President Lyndon B. Johnson signed the Civil Rights Act of 1964. And activism—and the need for it—didn't end there.

So: Scream, protest, debate, negotiate, cajole, tweet, use all the means at your disposal to call for the scale of policy change needed to match the magnitude of the climate challenge. To use the economists' logic of comparative advantage, do what you do best: Teachers, teach; students, study; community leaders, lead. Meanwhile, avoid *crowding-out bias* at every step and make sure to keep the next step in mind: the *Copenhagen Theory of Change* in action.

That's step #1. And it applies at every level—from city halls to state capitols to Washington, D.C., to capitals around the world, and to every level of the United Nations edifice. There are better and worse ways of screaming, but we won't pretend to know more than coteries of political strategists, focus-pollsters, and other hired guns.

Scream poorly, and it may backfire. Scream well, and it may yet pull us over the seemingly insurmountable legislative threshold.

For crying out loud, *scream well.*

STEP #2: COPE

Elisabeth Kübler-Ross gave us the five stages of grief. We are long past the time when denial, anger, bargaining, or depression is appropriate. The globe has already warmed by 0.8°C (1.4°F). Extreme weather looks to be the new normal. New York City was hit by two "100-year" storms within two years. The costs are piling up. The appropriate response: acceptance.

To be clear, we ought to do everything in our power to prevent further climatic changes. It's not a question of *if* we should set a price on carbon but how high it should be. It's clear that the optimal price is so high relative to where we are globally, that right now there's only one way to go: up. Increase the price on carbon. All that falls under: "Scream."

Coping with climate change comes with one all-important feature. Unlike doing something to prevent further climate change in the first place, coping is all about you. You buy an air-conditioner; you feel cooler—even though the planet warms ever so slightly as a result. That alone doesn't mean it's not the right thing to do for you individually, but there are better and worse ways to cope.

If you are committing to a thirty-year mortgage, think twice about that beachfront property. Any bank with sensible mortgage checks and balances may take one look at elevation maps and decide not to give you the loan in the first place. But don't trust that decision-making process. It's you who's left with the property after thirty years when sea levels have risen further, not the bank.

A better test may be where insurance premiums for climate-related risks are going. In most instances, they have nowhere to go but up. Lloyd's of London, Munich Re, and Swiss Re, the insurers of last resort who are left holding the ultimate risk, have been warning about climate risks for years. In the end, insurers and re-insurers will be OK. They'll charge higher premiums or stop selling particular policies altogether and manage to keep their profit margins intact.

As long as higher insurance premiums signal that you shouldn't be rebuilding in a flood zone, that's just as well. But often it's the public left footing the bill, sometimes directly. Part of the tens of billions of dollars in federal Sandy aid have gone to rebuilding properties just as they were before Hurricane Sandy swept them away. That's a federally subsidized disaster in the making. Much better to go with New York governor Andrew Cuomo's suggestion and use part of that money to buy private properties and convert them into public lands. The next big storm will inevitably require additional emergency aid to help those hit the hardest. Every aid disbursement inevitably also creates unintended consequences of subsidizing people living in flood zones. Governments can't abdicate their responsibility to help those worst hit, but they clearly ought to stop fueling this vicious circle.

While we shouldn't be giving aid to homeowners to rebuild their flooded homes on remote barrier islands, some adaptive actions in the face of rising seas may well be warranted. It's not a secret that eventually, we'll need to move much of New York City to higher grounds—unless, or possibly even if, we "scream" loudly enough. It's also no secret that in the meantime, putting up higher seawalls may be the best option and well worth the cost. The Dutch figured this out a long time ago. Their seawalls are necessary simply because large parts of the Netherlands are already below sea level, entirely without climate change. New York is now facing similar questions around building flood gates to prevent storm surges and worse from engulfing the city. What used to be a 1 percent chance of storm tides breaching typical New York seawalls in the mid-1800s has since risen to 20 to 25 percent per year. Manhattan is home to hundreds of billions of dollars' worth of real estate, all concentrated in a relatively small area. A seawall to prevent the worst may be comparatively cheap. That's what the Dutch do on a much broader scale.

|||||||||||||||||

Coping is all about planning ahead. If you have a Dutch friend living behind a dam, tell her to *Plan voor het ergste.* Plan for the worst. You buy insurance not in the hope of something going wrong, but to insure yourself in case it does. The overwhelming majority of fire insurance policies never pay any money. That's exactly how the insurance company can afford to sell you a policy in the first place.

No one knows whether the next 100-year flood will hit New York City next year or next decade. (We are pretty sure by now that it will be before next century.) All we know is that we can't get complacent. That same logic also goes for the longer run.

Patek Philippe is famous for its ad campaigns featuring a proud parent with his or her child. The fourth-generation family-owned watch company would like you to start your own family tradition, ideally by going out and buying one of their watches to then bestow onto your progeny's wrist. Some New York real estate companies have started to pick up on that with slogans like: "Own for generations." Whether or not that's actually possible may depend on how many generations you are thinking ahead. If your horizon ends with your grandkids, chances are you'll be fine. But it won't take all that many generations for some of these property owners in lower Manhattan to face the same choice some more remotely located New Yorkers (like those in Breezy Points, Queens) face now: rebuild after the flood, or just move to higher ground?

Whatever you do, hold on to that Canadian or Swedish or Russian citizenship. You and your grandkids may still want to vacation further south, but massive changes are upon us.

STEP #3: PROFIT

Imagine the *700 ppm Fund*. That number is far from set in stone, but it's the International Energy Agency's (IEA) best guess for where we are heading by 2100. IEA takes into account all countries' stated emissions reductions targets and then some. This already optimistic scenario would imply a 50 percent chance of global average warming of over 3.4°C (6.1°F) above preindustrial temperatures and about a 10 percent chance of eventual global warming exceeding 6°C (11°F). Think back to chapter 1: Mark Lynas's reference to Dante's Sixth Circle of Hell, or HELIX, the European Union research project detailing the impacts. Both Lynas and HELIX end their nightmare scenarios at 6°C (11°F). We are looking at a 1-in-10 chance of going there or beyond. It's tough not to sound overly dramatic about what that world would look like. Sea levels were up to 20 meters (66 feet) higher last time concentrations hit 400 ppm. A 700 ppm planet would look very different from anything we can imagine today. Still, that's our current trajectory.

You have $1 billion under management to invest in such a world. One way to profit is to invest in restoring the value of damaged assets: Someone has to be pumping away flood waters and rebuilding homes. The flipside of the cost of coping with climate change: *ka-ching!* Enormous costs imply large profit opportunities.

One crucial issue is the time scale involved. Many of the effects are decades out, though plenty are not. Extreme storms, droughts, and floods are already hitting home. Buy food staples, fresh water, or any kind of commodity that will be scarcer and, thus, dearer in a much warmer and more unstable world. To make a buck off the general trend, buy all of these assets while there are still plenty of climate skeptics around, who are betting against the general trend, to ensure you get in before prices really take off. And, since we are imagining investment opportunities in a world heading toward 700 ppm, buy stakes in mining and oil companies savvy enough to drill in the newly ice-free Arctic.

||||||||||||||||||

Now imagine the *350 ppm Fund*. We have long passed that threshold. Right now we are at 400 ppm for carbon dioxide alone. And the atmospheric tub is still filling at an increasing rate. Returning to 350 ppm would require an immediate about-face and then some. Everything we know about the economics tells us that this won't and can't happen. "Simply" turning off all smokestacks—an enormous, immediate, and global deindustrialization—will no longer do. If anything, it would require an enormous reindustrialization to transform energy and transport sectors *and* start capturing carbon directly from the air.

Lots of infrastructure will still be at risk for many years to come. We will have locked in decades, even centuries, of sea-level rise and plenty of unknown unknowns. It may already be too late for the West Antarctic ice sheet, but the world may steer clear from other tipping points with even costlier effects.

Stranded assets dominate the picture. Bill McKibben popularized the concept in *Rolling Stone*. The Capital Institute did the math for him: Just to stabilize atmospheric carbon dioxide concentrations at 450 ppm, about $20 trillion dollars' worth of carbon still underground will likely have to remain there or be pumped out only while pumping the resulting carbon dioxide back in, devaluing fossil fuel companies in the process.

In this world, your $1 billion may be best served betting against coal, oil, and gas. They are bound to perform worse than the broader market. Wind, solar, and all sorts of low-carbon technologies win. Carbon air capture technologies may be another big winner, assuming that the carbon dioxide price we all pay will be appropriately large. Once again, timing is everything. In order to make a buck, it will be key to get in at just the right time.

|||||||||||||||||||

The truth ought to be somewhere in the middle, between the business-as-usual nightmare of the 700 ppm and the green dream of the 350 ppm worlds. Just to be clear, there's an important difference between these two numbers: 700 ppm is where we *are* heading. Calling for "350 ppm" is a statement about where we *wish* to be heading. The two are in entirely different categories. There is hope that if we scream loudly enough and scream well, the world could maneuver away from the precipice of the 700 ppm future and toward an outcome closer to 350 ppm, but that's far from a certainty.

So, what to do with your hypothetical $1 billion? First you need to realize that smart investment decisions are all about what is (or, rather, what *will* be), not what *ought* to be. The current Arctic gold rush is an all-too clear example of that. Those not holding their noses and, thus, not participating in the bonanza of newly opened shipping lanes, mines, and oil fields may well be losing out.

That said, some recent evidence suggests that the more socially conscious of enterprises may, in fact, be outperforming the market, sometimes quite significantly. But our advice isn't about the fact that seeing things through green-tinged glasses could help identify opportunities that the market misses. Instead, we'll focus on the fact that smart investment decisions are all about managing risk. There's a distinction between risks to the planet and risks to Big Coal, Oil, and Gas, but there's also an important link: Regulations and policies for the most part point in only one direction.

Few doubt that, given trends in the regulatory climate, the arrow for tobacco stocks points downward. In Australia, tobacco companies are required to sell their products in plain packaging, with graphic health warnings. The 2011 Tobacco Plain Packaging Act was met with fierce resistance by a handful of tobacco companies, who stood to lose a lot and argued in court that the act was unconstitutional. The day the Australian High Court rejected these arguments and upheld the law in August 2012, British

American Tobacco and Imperial Tobacco share prices each dropped by about 2 percent. The High Court decision could have gone the other way, possibly lifting stock prices, but it's highly unlikely that governments will suddenly decide that tobacco has been vilified for too long and start removing packaging restrictions and smoking bans. If anything, more cities will follow Mayor Michael Bloomberg's New York and banish smokers onto the sidewalks, or worse. Investors concerned about managing risks ought to take note.

Something similar goes for anyone wanting to invest in Big Coal or Oil. Regulation, for the most part, will only push coal or oil company valuations down, not up. They are the ones with the stranded assets once we have a sensible price on carbon dioxide. It's highly unlikely that governments will suddenly start taxing wind and solar companies and increase subsidies for fossil fuels even further. (Big Gas may be in the gray zone: Initial regulations may price coal-powered generation out of the electricity system once and for all, making natural gas the fuel of the moment—at least until it, too, runs up against ever tightening greenhouse gas limits. It may be a "bridge" to a low carbon future, but that doesn't mean it wouldn't justify eventual heavy tolls in its own right.)

In short: Divest, because it's the prudent, less risky financial decision. It helps you hedge against downside regulatory risks and stranded assets. The move from the 700 ppm path toward a 350 ppm one will come in political fits and starts. Not investing in fossil fuel stocks is not just the ethical choice, it may well be the profitable one.

All that said, all-out divestment from fossil fuel stocks may not be the only ethical choice. Better yet: Apply that socially conscious screen to what you do with the returns.

Why cede the ground of (sadly) profitable investments to those without scruples and with no desire to influence the current trajectory?

It's clear that all of us—at least the billion or so high-emitters on this planet, including most anyone reading (or writing) this book—have been profiting from a world heading toward warmer climates all along. That doesn't make it right, but there's indeed an ethical path forward: Now that the reality of heading toward 700 ppm and the mandate of bending the path toward 350 ppm have become abundantly clear, take your outsized returns and make your money work even harder by helping scream for the biggest policy push your newly found wealth can muster.

A Different Kind of Optimism

The evidence is overwhelming: levels of greenhouse gases in the atmosphere are rising. Temperatures are going up. Springs are arriving earlier. Ice sheets are melting. Sea level is rising. The patterns of rainfall and drought are changing. Heat waves are getting worse as is extreme precipitation. The oceans are acidifying.

THESE WORDS COME FROM the American Association for the Advancement of Science (AAAS a.k.a. "triple-A, S"). The report is surprising only for the direct language it uses. None of these conclusions advances science itself. As the title of the report suggests, they simply describe "What We Know."

WHAT WE KNOW IS BAD, WHAT WE DON'T IS WORSE

Any homeowner would be well advised to fix the boiler in danger of overheating or the leaky valve at the gas stove, lest either end in catastrophe. But in addition, most home-owners take out fire insurance for the unlikely event that the entire house burns down in a freak accident.

That's not wishing for catastrophe to strike. It's not alarmism. It's the prudent move. In the unlikely event of a fire destroying the entire home the cost would be too large to skimp on the insurance premium.

Ironically, it's precisely those insurance premiums for catastrophic events like floods and droughts where the effects of climate change on our wallets may be hitting home the soonest. Federal flood insurance is heavily subsidized for all sorts of messy political reasons. It shouldn't be, as it encourages homeowners to build in particularly risky areas. And it won't take many more hurricanes hitting New York City to lead to necessary reforms of the entire system, which will make it that much more expensive to own a home in a flood zone.

Increasingly intense hurricanes, more floods, more droughts, not to say anything of rising temperatures and rising seas are *what we know is happening and will continue to happen*. Tallying those effects—at least the bits we can, in fact, put a dollar figure on—results in a minimum cost of $40 per ton of carbon dioxide we pump into the atmosphere today. But on average, the world isn't considering anything close to these costs. The average global price is closer to *negative* $15 per ton, considering the massive fossil fuel subsidies in many countries.

||||||||||||||||||

None of that yet includes the truly frightening low-probability events. There's a huge difference between a likely sea-level rise of 0.3 to 1 meters (1 to 3 feet) by the end of this century and eventual possible extremes of 20 meters (66 feet) or more in future centuries. And it's debatable whether we can describe any of these extreme scenarios as "unlikely" or "low probability" to begin with. By our own, conservative calculations, there's about a 1-in-10 chance of eventual global average warming in excess of 6°C (11°F), something that can be described only as "catastrophic" for society as we know it.

Any such talk of inevitability engenders calls of alarmism. It's anything but. We see it as our obligation to paint the full picture of what we know, and to show how what we don't know might play out. We take no satisfaction in doing so. We can only hope that we are wrong.

WRONG, THRICE OVER

First, we hope we are wrong in the sense that the really bad low-probability events will never come to fruition.

Second and more importantly, we hope we are wrong because society will manage to steer the climate ship away from the proverbial iceberg by severely cutting the flow of carbon into the atmosphere. There's plenty of warming and sea-level rise, and there are more floods, more droughts, and more of all sorts of other weather extremes already baked in, but rapid action can help deter the worst predictions.

Third, we hope we are wrong about the seemingly unstoppable drive toward geoengineering: shooting sulfur or other particles into the stratosphere to create an artificial sun shield. Everything we know about the economics tells us that the same fundamental forces that make it difficult to do much about climate change in the first place make it likely that we are going to face a geoengineered planet at some point, and possibly one geoengineered in some "rogue" fashion. The climate problem is too big, and it has too much momentum, while the geoengineering technology is too cheap and too readily available.

Our hope is that we will be proven wrong on all three counts. The world gets lucky on the science, works out the seemingly intractable politics of cutting emissions, and comes up with an iron-clad governing mechanism to guide geoengineering research into a productive direction and away from the seeming near-inevitability of (rogue) geoengineering.

STICK IT TO CARBON

It would be easy to conclude that economics—capitalism—is *the* problem. Economics is indeed at the core of the problem. Or rather: misguided market forces are.

One seeming solution then would be to simply change our ways. If only we slowed down, went back to the land, and generally did more with less, climate change would be a thing of the past. Not quite. Most would like to spend more time with our families frolicking in fields of green and less time tied to our desks. But that's clearly not enough. The math on voluntary action simply doesn't add up. And the calculus of changing capitalism as we know it—however desirable that may be as an independent goal—is daunting, to say the least. It also confuses the issue.

Some, like activist author Naomi Klein, call for "taxing the rich and filthy." That's a nice turn of phrase. One might agree that we probably should be taxing the rich more. But that's a different problem entirely. First and foremost, we ought to be taxing the filthy. Instead of "sticking it to the man," the point is to *stick it to carbon*.

Far from posing a fundamental problem to capitalism, it's capitalism with all its innovative and entrepreneurial powers that is our only hope of steering clear of the looming climate shock. That's not a call for letting markets run free. *Laissez-faire* may sound good with the right French accent—in theory. But it can't work in a situation where prices don't reflect the true costs of our actions. Unbridled human drive—erroneously bridled drive, really—is what has gotten us into this current predicament. Properly channeled human drive and ingenuity, guided by a high enough price on carbon to reflect its true cost to society, is our best hope for getting us out.

Only then can we afford the luxury of talking about what would truly be an ethical solution: for carbon pollution to go the way of child labor and slavery—something to be avoided on purely moral grounds. Kick out the economists, and call in the priests, imams, rabbis, or your favorite nondenominational philosopher. Just not quite yet. Moving to the moral high ground requires having high ground left that's not yet inundated by rising sea levels. That, in turn, requires taking the economics seriously.

Acknowledgments

||||||||||||||||||||||||||||||||||||

T HIS BOOK IS BASED ON around a dozen papers written over the course of a decade, many more ideas that we've adapted from others, and countless conversations trying to refine our logic and presentation.

We thank first and foremost our editor Seth Ditchik, who saw potential when all we had was an overly long collection of thoughts. Princeton University Press has proven to be a superb outlet for this book, enabling us to gain invaluable insights from three anonymous peer reviewers, while at the same time allowing us to brush off any suggestion that we ought to add equations to a text that includes none. (For the occasional equation as well as in-depth discussions and detailed references, see our Notes.)

Peter Edidin and Eric Pooley helped shape the ideas before we had written the first word. Liza Henshaw made it all possible. Rob Socolow helped us portray his quiz in the preface in the right light. Dorothy Barr at Harvard's Ernst Mayr Library worked tirelessly to confirm our lines around "camels in Canada" from peer-reviewed sources. Bob Litterman provided valuable insights on the theory and practice of asset pricing and pointed us to the Sir Richard Branson quote on a global carbon tax being "blindingly obvious." Many others provided invaluable insights, comments, and discussions, including Richie Ahuja, Joe Aldy, Jon Anda, Ken Arrow, Michael Aziz, Len Baker, Scott Barrett, Seth Baum, Eric Beinhocker, Jennifer Chen, Frank Convery, Kent Daniel, Sebastian Eastham, Denny Ellerman, Ken Gillingham, Timo Göschl, Steve Hamburg, Sol Hsiang, Matt Kahn, David Keith, Bob Keohane, Nat

Keohane, Matt Kotchen, Derek Lemoine, Kathy Lin, Frank Loy, Charles C. Mann, Michael Mastrandrea, Graham McCahan, Kyle Meng, Gib Metcalf, George Miller, Juan Moreno-Cruz, David Morrow, Bill Nordhaus, Ilissa Ocko, Michael Oppenheimer, Richard Oram, Bob Pindyck, Billy Pizer, Stefan Rahmstorf, Colin Rowan, Dan Schrag, Jordan Smith, Rob Stavins, Elizabeth Stein, Thomas Sterner, Cass Sunstein, Claire Swingle, Johannes Urpelainen, David Victor, Jeff Vincent, Matthew Zaragoza-Watkins, and Richard Zeckhauser.

Katherine Rittenhouse provided invaluable research assistance every step along the way. Katherine, Keith Gaby, Peter Goldmark, and Tom Olson read every word and many more that, thanks to them, didn't make it into this final version.

None of this would have been possible without Siri Nippita and Jennifer Weitzman, who helped with everything from reading early drafts to indulging our lengthy book conversations during late-night calls as well as over occasional Sunday and holiday brunches. Writing *Climate Shock*—much like the looming climate shock itself—has proven to be all-absorbing at times. And in the end, unknowns may well prevail: All remaining errors are our own. The same goes for our views. They are ours and ours alone. They should not be attributed to anyone credited here, to the Environmental Defense Fund, the Trustees of Columbia University, the President and Fellows of Harvard College, or anyone else with whom we are, ever have been, or wished we were affiliated.

Notes
||||||||||||

PREFACE: POP QUIZ

Page ix—**Two quick questions:** Princeton's Robert Socolow has started many a presentation with a version of this quiz, asking audiences whether they consider climate change "an urgent matter" and fossil fuels "hard to displace." He groups the resulting views into four broad buckets, reproduced here with permission, and with slight modifications:

		Is getting the world off fossil fuels difficult?	
		No	Yes
Is climate change an urgent problem?	No	A low-carbon world unmotivated by climate considerations.	Perhaps most of the general public, and parts of the energy industry.
	Yes	Many environmentalists, including nuclear advocates.	Our working assumption.

Socolow, "Truths," searches for solutions firmly grounded in this "working assumption." Oliver Morton, editor at the *Economist*, introduced an August 2013 debate on geoengineering at the Massachusetts Institute of Technology with these two questions. Morton echoed Socolow's conclusion that, to avoid cognitive dissonance, most people answer "Yes" to either one or the other question, but not both. In the packed lecture hall that evening at MIT, most answered "Yes" to both, a clear indication of the type of people currently attracted to geoengineering conversations.

Page x—**Standard economic treatments:** For a popular, standard perspective on the science and economics of climate change, see Nordhaus, *Climate Casino*. For more, see the "DICE" entry on page 36 in chapter 2.

CHAPTER 1. 911

Page 1—**exploded in the sky**: See Artemieva, "Solar System: Russian Skyfall."

Page 1—**$2 or 3 billion**: Section 321 of the NASA Authorization Act of 2005 directs NASA both to "detect, track, catalog and characterize certain near-earth asteroids and comets" and to write a report including "analysis of possible alternatives that NASA could employ to divert an object on a likely collision course with Earth." The options range from "non-nuclear kinetic impactors," described as the most mature technology, to a "nuclear standoff explosion," possibly the most effective ("Near-Earth Object Survey"). See "Defending Planet Earth" about the inadequacy of current funding. The report concludes that $250 million per year for ten years would allow NASA to launch an actual test of deflecting an asteroid.

The United Nations recognizes asteroid deflection as a global issue and recently voted to create the "International Asteroid Warning Group," where members will share information about potentially dangerous approaching asteroids, and work with the UN Committee on the Peaceful Uses of Outer Space to launch a defense. The UN began discussing the creation of such an international warning group after a meteor exploded over Russia in February 2013, without the world's space agencies knowing beforehand ("Threat of Space Objects Demands International Coordination, UN Team Says").

Page 1—**1-in-1,000-year event**: An asteroid impact of the size that may warrant a full-on defense may be a one-in-1,000-year event. The probability of an asteroid impact the size of the one that exploded above Chelyabinsk Oblastin February 2013 is commonly seen to be around one in 100 years (Artemieva, "Solar System: Russian Skyfall"). However, the latest research puts the probability of a Chelyabinsk-sized asteroid at ten times that estimate (Brown et al., "500-Kiloton Airburst").

Page 2—**major extinction event**: Kolbert, *Sixth Extinction*, looks at prior extinction events and then mainly focuses on the current, human-caused one. For a summary of Kolbert's arguments, see: Dreifus, "Chasing the Biggest Story on Earth."

Page 2—**past 65 million years**: Diffenbaugh and Field, "Changes in Ecologically Critical Terrestrial Climate Conditions." This even includes the Palaeocene–Eocene Thermal Maximum (PETM) around 56 million years ago, when the globe warmed by at least 5°C (9°F) in less than 10,000 years, a rate of change still ten times slower than the projected rate of global average surface temperature increase in the IPCC's RCP2.6 scenario.

Page 2—**100-year flood**: Lovett, "Gov. Cuomo."

Page 2—**Irene killed 49**: Avila and Cangialosi, *Tropical Cyclone Report*. "Irene by the Numbers" estimates that 2.3 million people were under evacuation orders in the United States.

Page 2—**Sandy killed 147**: Blake et al., Tropical Cyclone Report.

Page 2—**Typhoon Haiyan**: As of January 28, 2014, Haiyan was estimated to have displaced 4.1 million, and killed over 6,000 people ("Philippines: Typhoon Haiyan Situation Report No. 34"). In the Philippines, Haiyan was named "Typhoon Yolanda."

Any of these figures are likely significant underestimates, as they exclude estimates of the negative impact storms have on a family's ability to properly care for themselves and their children. Antilla-Hughes and Hsiang, "Destruction, Disinvestment, and Death," shows that "unearned income and excess infant mortality in the year after typhoon exposure outnumber immediate damages and death tolls roughly 15-to-1."

Page 2—**Typhoon Bopha**: According to "Report: The After Action Review / Lessons Learned Workshops for Typhoon Bopha Response," Typhoon Bopha affected 6.2 million people, destroyed 230,000 homes, and killed 1,146 people, with another 834 still missing. The latest situation report published on the effects of Bopha counts over 700,000 who have sought shelter in evacuation centers plus 1.06 million outside evacuation centers during the peak of displacement, for a total of 1.76 million displaced (National Disaster Risk Reduction and Management Council), rounded to 1.8 million in the text. Also see Antilla-Hughes and Hsiang, "Destruction, Disinvestment, and Death," for why these numbers are likely significant underestimates of full costs and deaths.

Page 3—**European summer heat wave**: Robine et al., "Death Toll."

Page 3—**equipped to cope**: Deschênes and Moretti, "Extreme Weather Events," estimate that Americans' mobility from the Northeast to the warmer Southwest climate has significantly increased average life expectancy since 1980. Barreca et al., "Adapting to Climate Change," highlights the importance of residential air-conditioning in the drastically declining temperature-mortality relationship in the United States.

Page 3—**waters off the coast**: Tollefson, "Hurricane Sandy," discusses the link between climate change and hurricanes. It also notes that "the expected increase due to global warming is just 0.6°C," concluding that "while the changing climate certainly plays a part . . . there is plenty of space for natural variability." Pun, Lin, and Lo, "Tropical Cyclone Heat Potential," discusses the recent warming trends in the water east of the Philippines, which most likely contributed to the severity of Typhoon Haiyan. Normile, "Supertyphoon's Ferocity," draws this link.

By comparison, global average sea surface warming has been around 0.1°C (0.2°F) per decade in the past four decades (Summary for Policymakers of Working Group I of the *IPCC Fifth Assessment Report*).

Page 3—**more** *and* **bigger storms:** Emanuel, "Increasing Destructiveness," published in 2005, showed that hurricanes had intensified over the preceding three decades. The ensuing scientific debate seems to have settled with the conclusion that climate change does indeed lead to more intense hurricanes but that their frequency may not change (or may even go down slightly). Some of the latest research, Emanuel, "Downscaling CMIP5," finds that climate change will likely lead to both more intense *and* more frequent storms. That scientific debate isn't settled, yet the physical signs are sadly clear. Projected economic impacts are similarly striking: Mendelsohn et al., "Impact of Climate Change," finds that "global hurricane damage will about double owing to demographic trends, and double again because of climate change" through 2100 (Emanuel, "MIT Climate Scientist Responds").

That said, hurricanes are still among the most difficult weather events to link to climate change, largely because of their rarity. As our ability to forecast hurricanes improves, it will become increasingly easier to conduct event studies around hurricanes of the type already conducted for other extreme events. (See the following note on "attribution science.")

Page 4—**attribution science:** A good starting point is the IPCC's 2012 *Special Report: Managing the Risks of Extreme Events and Disasters to Advance Climate Change Adaptation.* The study reports mixed evidence around today's extreme events but increasingly certain evidence going forward. There are also increasingly detailed studies of single events, perhaps most prominently by Peter Stott, who leads the Climate Monitoring and Attribution team at the UK Met Office. Stott, Stone, and Allen, "Human Contribution," draw the conclusions mentioned in the text, a doubling of the risk of a heat wave of the magnitude observed in Europe in 2003. Stott et al., "Attribution of Weather," surveys the recent literature and points to a way forward for attribution science.

A slew of other papers, many from the Met Office's Climate Monitoring and Attribution team, highlight the contributions of the rapidly developing field of "attribution science": Christidis et al., "HadGEM3-A Based System for Attribution," finds that the 2010 heat wave in Moscow can be attributed, at least in part, to human-caused climate change. The study compares a model run with observational data with estimates of what those data would be without anthropogenic forcings. Rahmstorf and Coumou, "Increase of Extreme Events," develop a method to determine the effect long-term trends have on the number of climate extremes. They use their approach to estimate that there is an 80 percent chance that the 2010 Moscow heat record would not have occurred without climate change. Otto et al., "Reconciling Two Approaches," contrasts these findings around increased odds with another study that finds no human fingerprint on the magnitude of the Moscow heat wave. Lott, Christidis, and Scott, "East African Drought," find that anthropogenic forcings increased the probability of the 2011 East African drought. Pall et al., "Flood Risk," uses a "probabilistic event attribution framework" to find that human emissions increased the likelihood of the 2000 floods in England and Wales by anywhere from 20 percent to over 90 percent. Peterson, Stott, and Herring, "Explaining Extreme Events of 2011," use the Central England Temperature dataset and global climate models to look at the effect of anthropogenic forcings on the chances of six extreme events in the UK that year. Li et al., "Urbanization Signals," attributes differences in minimum winter temperatures in Northern China cities to urbanization effects.

Some others look at the global links between warming and extreme events. Coumou, Robinson, and Rahmstorf, "Global Increase," look at the rising likelihood of record-breaking monthly-mean temperatures due to climate change. Their verdict: "Under a medium global warming scenario, by the 2040s we predict the number of monthly heat records globally to be more than 12 times as high as in a climate with no long-term warming." Also see Coumou and Robinson, "Historic and Future Increase," which estimates the percent of global land area that can expect to experience extreme summer heat.

Page 5—**three to twenty years**: See Rosenzweig and Solecki, "Climate Risk Information," and Fischetti, "Drastic Action." Lin et al., "Physically Based Assessment," uses a combination of climate and hydrodynamic models to show that what are now 100-year floods may hit every three to 20 years by the end of the century.

Talke, Orton, and Jay, "Increasing Storm Tides," estimates the increased chance of annual seawall breaches today compared to the mid-1800s. Also see Kemp and Horton, "Historical Hurricane

Flooding," who look at the direct contribution of sea-level rise to hurricane flooding.

Page 5—**0.3 to 1 meters**: The range comes from Working Group I's Summary for Policymakers in the *IPCC Fifth Assessment Report*. It compares average global sea levels for 2081–2100 to 1986–2005. That number is significantly above the earlier estimates from the *IPCC Fourth Assessment Report*. (See "left it out" on page 11.) It also updates (and lowers) earlier high estimates by the U.S. Army Corps of Engineers, which operates with a high scenario of 1.5 meters (5 feet) ("Incorporating Sea-Level Change Considerations in Civil Works Programs"), and by the National Oceanic and Atmospheric Administration (NOAA), which uses 2 meters (6.6 feet) as its high scenario for 2100 (Parris et al., "Global Sea Level Rise Scenarios").

Page 5—**no replacement**: Gillett et al., "Ongoing Climate Change," argues that if "rapid melting of the West Antarctic ice sheet . . . were driven by ocean warming at intermediate depths, as is thought likely, a geoengineering response would be ineffective for several centuries owing to the long delay associated with subsurface ocean warming." A full melting of the West Antarctic ice sheet would lead to about 3.3 meters (11 feet) of sea-level rise. (See "Melting of Greenland" on page 56 in chapter 3.)

Page 6—**catastrophe**: Kolbert, *Field Notes from a Catastrophe*, is among the most eloquent accounts. For a seminal study on the definitions of "dangerous anthropogenic interference," see Ramanathan and Feng, "Avoiding Dangerous Anthropogenic Interference." For a seminal classification of "tipping elements in the earth's climate system," not all of them necessarily "catastrophic," see Lenton et al.'s eponymous study. The list includes the melting of Arctic summer sea ice, the melting of the Greenland ice sheet, the melting of the West Antarctic ice sheet, the shutoff of the Atlantic thermohaline circulation, increasingly intense El Niño / Southern Oscillation, changes to the Indian summer monsoon, and dieback of the Amazon rainforest. Assessments differ for many of these potential tipping elements, which makes their potential impacts all the more significant. (See our extensive discussion of uncertainty in chapter 3.)

Page 8—**"free-driver" effect**: See the "Free Drivers" entry on page 38 in chapter 2 for a more comprehensive definition. See the corresponding note for alternative uses in the academic literature on energy efficiency economics, where it describes a type of network effect.

Page 8—**Chinese soot**: Bradsher and Barboza, "Pollution from Chinese Coal"; Yienger et al., "Episodic Nature."

Page 9—**warmest in human history**: Despite these clear overall trends, some have invariably pointed to a so-called warming pause or hiatus this past decade, which has found resonance in the press. See, for example: Ogburn, "What's in a Name?," Ogburn, "Climate Change 'Pause' into Mainstream," and Voosen, "Provoked Scientists." For a comprehensive analysis of media coverage, see Greenberg, Robbins, and Theel, "Media Sowed Doubt." The latest research points to the fact that the drop in the rate of warming wasn't there in the first place, providing a number of insights that, put together, may even over-explain the drop ("Global Warming: Who Pressed the Pause Button").

Page 9—**National Climate Assessment**: See Melillo, Richmond, and Yohe, "Climate Change Impacts in the United States."

Page 9—**"The Coming Arctic Boom"**: Borgerson, "The Coming Arctic Boom."

Page 9—**decades of warming**: If greenhouse gas concentrations already in the atmosphere had been held at 2000 levels, we would still have been committed to a likely temperature rise of 0.3–0.9°C (0.6–1.6°F) by 2100 relative to 2000, with a best estimate of 0.6°C (1.1°F). That number comes from the *IPCC Fourth Assessment Report*, also cited in chapter 12 of the *IPCC Fifth Assessment Report*.

A complete halt in emissions would only very slowly decrease global temperature. Ramanathan and Feng, "Avoiding Dangerous Anthropogenic Interference," reviews work that shows how only about a quarter of the already baked-in global average warming has been realized so far. Coumou and Robinson, "Historic and Future Increase," finds that if we stopped emitting today, we would still be locked into a doubling of land area experiencing extreme summer heat by 2020, and a quadrupling of that area by 2040. Only after 2040 will the frequency and severity of heat waves depend greatly on our level of mitigation today.

Even "air capture" of carbon dioxide—taking carbon dioxide out of the atmosphere directly—has a considerable lag. Air capture, once implemented at scale, can slow the rate of further changes, but many of the intervening climatic changes will indeed be irreversible. (See "comes under various guises" on page 107 in chapter 5 as well as the "Bathtub" entry on page 30 in chapter 2.)

Page 9—**centuries of sea-level rise**: Meehl et al., "Relative Outcomes," finds that even under aggressive mitigation scenarios that stabilize temperatures, "sea-level rise cannot be stopped for at least the next several hundred years."

Two independent studies point to the eventual collapse of large parts of the West Antarctic ice sheet (Joughin, Smith, and Medley,

"Marine Ice Sheet Collapse," and Rignot et al., "Widespread, Rapid Grounding Line Retreat"). It has already been clear that the West Antarctic ice sheet has been melting at an ever increasing rate. Shepherd et al., "A Reconciled Estimate," estimate the average yearly loss of mass in the West Antarctic ice sheet to be 38 billion tons from 1992 to 2000, 49 billion tons from 1993 to 2003, 85 billion tons from 2000 to 2011, and 102 billion tons from 2005 to 2010.

Also see "Melting of Greenland" on page 56 in chapter 3.)

Page 10—**excess carbon dioxide in the atmosphere**: Solomon et al., "Irreversible Climate Change." Results differ across scenarios, but a rough rule of thumb suggests that approximately 70 percent of the "peak enhancement level" over the preindustrial level of 280 ppm perseveres after 100 years of zero emissions, while approximately 40 percent of the peak increase over the preindustrial level of 280 ppm persevered after 1,000 years of zero emissions. Note that this refers to the net increase in carbon dioxide in the atmosphere, not the exact molecule. Archer et al., "Atmospheric Lifetime," discusses the two often confused definitions for carbon's "lifetime," and concludes that 20–40 percent of excess carbon levels remain hundreds to thousands of years ("2–20 centuries") after it is emitted. The oft-cited Bern Model calculates that 20 percent of carbon dioxide remains after 1,000 years (Joos and Bruno, "Short Description"). The latest IPCC consensus says that roughly 15 to 40 percent of excess carbon dioxide remains in the atmosphere for over 1,000 years (see the *IPCC Fifth Assessment Report* Working Group I's Summary for Policymakers). Each carbon dioxide molecule has a lifetime of anywhere between 50 to 200 years, according to the U.S. Environmental Protection Agency's "Overview of Greenhouse Gases: Carbon Dioxide Emissions." The precise number is under considerable scientific dispute and is surprisingly poorly understood (Inman, "Carbon Is Forever").

Page 10—**400 parts per million**: 400 ppm is the concentration of carbon dioxide. Concentrations of other greenhouse gases—including methane, nitrous oxide, and industrial gases—are well known, too, but converting them into carbon dioxide–equivalent terms is fraught with uncertainties, as it relies on a number of assumptions of relative radiative efficiency compared to carbon dioxide and the atmospheric lifetimes of the gases over time. Estimates of carbon dioxide–equivalent concentrations range from around 440 to as high as 480 ppm ("World Energy Outlook 2013," citing a 2010 estimate, and Butler and Montzka, "NOAA Annual Greenhouse Gas Index," citing a 2013 estimate, respectively). See also Monastersky, "Global

Carbon Dioxide Levels," for a more detailed account of reaching the 400 ppm milestone.

Adding in the relative cooling effects of various tiny human-made particles (aerosols), brings total global warming effects of all human-caused emissions down to closer to around 400 ppm. Hence the best proxy for the full effect of all human-caused emissions *today* is still around 400 ppm, though if and when the masking effect of cooling aerosols disappears, the impact is bound to rise—and perhaps dramatically so.

The difficulties around converting everything into carbon dioxide–equivalent metrics is one reason why the IPCC primarily presents the warming impacts of human-caused emissions in terms of radiative forcing. The *IPCC Fifth Assessment Report* Working Group I's Summary for Policymakers puts total human-caused radiative forcing relative to 1750 at about 2.29 W m^{-2}, a level that includes the negative 0.9 W m^{-2} forcing from aerosols.

Page 10—**Global average temperatures:** Chapter 5, "Information from Paleoclimate Archives," of the *IPCC Fifth Assessment Report* lays out these facts about the Pliocene environment. Temperatures were 2–3.5°C (3.6–6.3°F) above preindustrial levels.

Page 10—**camels lived in Canada:** Rybczynski et al., "Mid-Pliocene" reports evidence showing the existence of giant camels living in the Canadian High Arctic in the Pliocene era.

Page 10—**decades to centuries:** The technical distinction is between a so-called fast equilibrium and so-called earth system sensitivity. Although time here is relative: "Fast" applies in geological terms, over the course of decades and even a century or two. Over the course of centuries, other factors that influence the earth's reaction to higher atmospheric concentrations of carbon dioxide begin to play a role. Examples include albedo changes, changes in biological sinks such as oceans and terrestrial ecosystems, and temperature-induced releases of carbon and methane. See, for example: Hansen et al., "Target Atmospheric CO$_2$," and Hansen and Sato, "Climate Sensitivity." Previdi et al., "Climate Sensitivity in the Anthropocene," incorporates these long-term feedbacks into an estimate of earth system sensitivity and finds that it could be twice as high as estimates for climate sensitivity, at 6–8°C (11–14°F) per doubling of carbon dioxide. Although this extra warming would be on a much larger timescale, perhaps multiple millennia, the effects of some of the feedbacks could start to hit home within this century.

Page 11—**third of sea-level rise:** The observed sea-level rise from thermal expansion since 1993 has been about 1.1 mm per year, or

34 percent of the total observed rise of approximately 3.2 mm per year. The modeled contribution of thermal expansion is higher, at 1.49 mm per year since 1993. Chapter 13, "Sea Level Change," of the *IPCC Fifth Assessment Report*.

Page 11—**left it out**: The 2007 IPCC report included the effects only from thermal expansion in its projections for sea-level rise, not the effects of melting polar ice caps (Projections of Future Changes in Climate in Working Group I of the *IPCC Fourth Assessment Report*), an omission since corrected. The Summary for Policymakers in Working Group I of the 2013 *IPCC Fifth Assessment Report* has several scenarios for sea levels, all incorporating the melting ice caps that were left out in the 2007 estimates, and projects sea-level rise as high as 1 meter (3 feet) by 2100 without significant climate action. For a good account of the debate surrounding the latest IPCC report and this particular issue, see Clark, "What Climate Scientists Talk about Now." See also "0.3 to 1 meters" on page 5.

Page 11—**perhaps even pleasant**: Moderate warming may indeed come with real, monetizable benefits. Virtually alone among climate-economic models, Richard Tol's FUND model estimates positive global benefits for slow, moderate warming up to about 2°C (3.6°F). For much of the 20th century, Tol estimates, the benefits of global warming may have outweighed the costs (Tol, "Economic Impact of Climate Change"). For another take around the opportunities provided by a changing climate, see Kahn, *Climatopolis*.

The broader question around the economic costs and benefits of global warming engenders considerable—often extremely contentious—debate. Tol, "Correction and Update," surveys 21 estimates of the welfare impacts of various degrees of average global warming. Three of these estimates, most notably Tol's own ("Estimates of the Damage Costs"), show zero or positive economic impact of climate change. (Tol, "Estimates of the Damage Costs," estimates a significant positive welfare impact of 1°C (1.8°F) of global average warming to the tune of 2.3 percent of global welfare. Mendelsohn et al., "Country-Specific Market Impacts," presents two central welfare estimates for 2.5°C (4.5°F) of global average warming, with both showing close to zero impact.) One further estimate, with a negative central value, spans zero as part of its confidence interval. The other 17 estimates surveyed show economic costs, some of them significant, at various global average temperatures. Tol then proceeds to plot all 21 economic impacts and presents the central "least squares" curve, including 95 percent confidence intervals ("Correction and Update," figure 2). Revising his own earlier estimates (reproduced in "Correction and Update," figure 1), Tol estimates that the central curve showing global welfare impacts is negative for any amount of global average warming. Even the upper 95 percent confidence interval barely goes above zero, a clear departure from Tol's own earlier survey updated and corrected here.

We would also hasten to add that most every one of the 21 estimates presented in Tol, "Correction and Update," can represent only a lower bound for the true economic costs. See "possibly much more" on page 23 in chapter 1 and our extensive discussion throughout chapter 3, in particular around "$2 per ton" and "Nordhaus's preferred "optimal" estimate" on page 57 as well as "damages affect output growth *rates*" on page 63.

Significant negative effects on human society and ecosystems notwithstanding, adapting to low levels of global average warming is a broad phenomenon. That may even include coral reefs, often viewed as a poster child of negative impacts: many fish will migrate; corals, by and large, can't. The newest evidence instead points to coping mechanisms for some corals (Palumbi et al., "Reef Coral Resistance"). Even while coping with warmer temperatures, however, marine environments still have to deal with the detrimental effects from increased acidity. See "Ocean acidification" on page 42 in chapter 2.

Page 12—**increasing rate**: Chapter 2 in Working Group I of the *IPCC Fifth Assessment Report* finds that global mean surface temperature has increased by approximately 0.86°C (1.5°F) since 1901, with 0.72°C (1.3°F) or 81 percent of that warming occurring since 1951. The reported average from 1951 to 2012 is 0.106 to 0.124°C (0.19 to 0.22°F) per decade, while the 100-year average, from 1901 to 2012, is only 0.075 to 0.083°C (0.14 to 0.15°F) per decade, depending on the dataset used. The U.S. Environmental Protection Agency, "Climate Change Indicators in the United States: U.S. and Global Temperature," shows that since the 1970s, the rate of increase has been 0.17–0.25°C (0.31–0.45°F) per decade in the United States, versus 0.072°C (0.13°F) per decade since 1901.

Something similar holds for sea-level rise. Sea levels have risen by around 0.2 meters (0.7 feet) over the past century. And the trend has been accelerating: over the past hundred years, the average sea-level rise was at around 1.7 centimeters (0.7 inches) per decade; over the past forty years, it was around 2.0 centimeters (0.8 inches) per decade; over the past twenty years, it was about 3.2 centimeters (1.3 inches) per decade. That trend is likely only going to accelerate further for the foreseeable future. The IPCC estimate for 2100 ranges from 0.3 to 1 meters (1 to 3 feet) in average global sea-level rise is relative to today's levels, on top of the 0.2 meters (0.7 feet) already observed. See "0.3 to 1 meters" on page 5.

Page 12—**decade without warming**: See "warmest in human history" on page 9.

Page 12—**increase over land**: From 2000 to 2009, temperature change in the US has been 50 percent greater over land than over ocean (Carlowicz, "World of Change"). Globally, surface air temperatures over land are thought to have risen 0.25 to 0.27°C per decade since 1979, depending on the dataset used, versus only 0.12°C per decade over oceans (chapter 2 in Working Group I of the *IPCC Fifth Assessment Report*).

Page 12—**twice the global average**: Average warming over the Arctic is projected to be 2.2 to 2.4 times the global average through the end of the century (chapter 12 in Working Group I of the *IPCC Fifth Assessment Report*).

Page 13—**warmed by 0.8°C (1.4°F)**: The *IPCC Fifth Assessment Report* Working Group I's Summary for Policymakers states two central numbers: 0.85°C of global average surface temperature warming between 1880 and 2012, and 0.78°C between the averages from 1850 to 1900 and 2003 to 2012. The 90 percent confidence intervals around each are 0.65 to 1.06°C and 0.72 to 0.85°C, respectively.

Page 14—**700 ppm**: The IEA's "World Energy Outlook 2014" calls this scenario the "New Policies Scenario." If we follow this trajectory, where all current emissions reductions commitments are met, support for renewables deployment and energy efficiency measures continues at or near current levels, and the world phases out at least parts of its fossil fuel subsidies, we can expect that carbon dioxide–equivalent concentrations reach 700 ppm by 2100. The IEA translates that level to a total temperature increase from preindustrial levels of 3.6°C (6.5°F), a bit more than our median increase of 3.4°C (6.1°F).

The IPCC is much less committal as to where concentrations will go. *IPCC Special Report on Emissions Scenarios* creates four families of scenarios, for a total of 40 cases, based on different sets of assumptions about the way the future world will work. It does not assign a probability to any of the scenarios and makes no claims about their relative likelihood. In later Assessment Reports, the IPCC uses these scenarios to determine ranges of possible future greenhouse gas concentrations. Frighteningly, the scenarios lead to estimates up to 1,550 ppm of carbon dioxide–equivalent concentrations. The latest IPCC report isn't any more reassuring. Its modeled scenarios range from a peak at 500 ppm to 1,500 ppm, with likely associated temperature increases this century of between 0.3 and 4.8°C (*IPCC Fifth Assessment Report* Working Group I's Summary for Policymakers).

Page 14—**Mark Lynas**: Lynas, *Six Degrees*. He describes in frightening detail what kind of changes we can expect with temperature increases of 1–6°C, starting with the loss of coral reefs and ending with extreme resource shortages and mass migration.

Page 14—**HELIX**: Short for: High-End cLimate Impacts and eXtremes. The project began in November 2013. For more, see www.HELIX climate.eu. The project description aims to provide "a set of credible, coherent, global and regional views of different worlds at 4, 6 and 2 degrees celsius."

Page 14—**around 10 percent**: See our discussion in chapter 3, in particular "clearly more room" on page 51 and "scientific papers" on page 53.

Page 14—**cognitive dissonance**: For the earliest work on cognitive dissonance and related phenomena, see Kahneman and Tversky, "Subjective Probability," Kahneman and Tversky, "Prospect Theory," and Kahneman, Knetsch, and Thaler, "Experimental Tests." Kahneman, *Thinking, Fast and Slow*, provides a comprehensive, accessible version, including its implications. For more on the psychology of climate change, see, among many others: Wagner and Zeckhauser, "Climate Policy."

Page 15—**giant bathtub**: Guy et al., "Comparing the Atmosphere to a Bathtub," conducted a study of the effectiveness of the bathtub analogy on increased understanding of carbon dioxide stabilization and one's preferred level of climate change mitigation. They found that the analogy can be effective in improving nonexperts' understanding of climate change. (They tested both undergraduate students and the Australian public.) The study also showed that using the analogy to explain carbon dioxide accumulation could lead to stronger support for climate action (in their test among undergraduate students). There is plenty of nuance, though. Words seem to help; graphs don't: "Our results show that analogy can improve non-experts' understanding of CO_2 accumulation but that using graphs to convey emissions rate information is detrimental to such improvements." For more on the bathtub analogy, see the "Bathtub" entry on page 30 in chapter 2.

Page 15—**More specifically**: Sterman, "Risk Communication." The specific question involved two graphs: One graph was given to test subjects and included a flat line for concentrations: "Consider a scenario in which the concentration of CO_2 in the atmosphere gradually rises to 400 ppm, about 8 percent higher than the level in 2000, then stabilizes by the year 2100." The second graph showed a rising trend line for emissions, asking students to fill in the future path of emissions to achieve stable concentrations. A surprising number of test subjects answered by stabilizing emissions instead of bringing them down to stabilize concentrations.

Page 15—**that won't happen**: Overall, net global uptake has doubled from about 8.8 to 18 billion tons of carbon dioxide per year between 1960 and 2010 (Ballantyne et al., "Increase in Observed Net Carbon Dioxide Uptake"). That comes out to about 50 percent of carbon dioxide emitted each year. In other words, the drain has increased because of the higher water pressure, even though the water is still rising. More recently, however, the rate of increase of uptake by oceans seems to have decreased, possibly hinting at a saturation point (Khatiwala, Primeau, and Hall, "Reconstruction of the History"). The same seems to hold for European forests (Nabuurs et al., "First Signs"). Reichstein et al., "Climate Extremes," points to significant caveats going forward.

Page 16—**latest IPCC report at the time**: See the Executive Summary of Working Group I in the *IPCC Fourth Assessment Report*.

Page 17—**declined by 80 percent**: Liebreich, "Global Trends." Lower manufacturing costs, rather than shorter-term stock liquidation, has caused much of the recent photovoltaic price reductions (Bazilian et al., "Re-considering the Economics of Photovoltaic Power").

Page 18—**50 percent of its electricity**: Kirschbaum, "Germany Sets New Solar Power Record."

Page 18—**5 percent of its electricity**: Photovoltaics accounted for 5 percent of total power consumption in Germany in 2013 (Franke, "Analysis"). It similarly accounted for 4.7 percent of total power consumption in 2013 ("Statistic Data on the German Solar Power [Photovoltaic] Industry").

Page 18—**looking up globally**: "China's 12GW Solar Market Outstripped All Expectations in 2013," and "Global Market Outlook for Photovoltaics 2013–2017." An important caveat here is that the impressive solar capacity growth masks relatively small capacity factors compared to traditional energy sources like fossil, nuclear, and hydro. Still—in keeping with the "optimism track"—increased generation follows a similar trend as increased capacity, and capacity factors are only going to improve going forward.

Page 19—**majority of the electorate**: It's true that the polls show general skepticism about the existence of climate change (see Marlon, Leiserowitz, and Feinberg, "Perspectives on Climate Change," where

97 percent of climate scientists versus 41 percent of the American public believe climate change is happening and caused by humans). Not surprisingly, Americans seem to be against many potential governmental actions to mitigate global warming, like a carbon tax (*Survey Findings on Energy and the Economy*). However, there are many pro-environment actions that a large majority (as high as 75 to 85 percent) of Americans polled would like to see. See Krosnick, "The Climate Majority," where data from the Political Psychology Research Group's poll of 2010 shows that Americans overwhelmingly support limitations on businesses' air pollution (86 percent), incentives or regulations to increase manufacturing of cars that use less gas (81 percent), appliances that use less electricity (80 percent), and buildings requiring less energy to heat and cool (80 percent). Moreover, young people overwhelmingly support climate legislation, according to a recent poll by the League of Conservation Voters (Benenson Strategy Group and GS Strategy Group). Eighty percent of voters under 35 support the president taking action on climate change. Over half of under-35 Republican voters would be less likely to vote for someone who opposed the President's Climate Action Plan. Finally, according to a Pew Research Center / USA Today survey, 62 percent of Americans are in favor of stricter emission limits on power plants. Americans on the whole are, however, less concerned about climate change than other countries surveyed by the Pew Research Center in the Pew Global Attitudes Project. Only 40 percent of Americans see global climate change as a major threat to their country. The global average for the 39 countries surveyed was 54 percent, the same as the percentage of Europeans who see climate change as a threat.

Page 19—**Technology is good**: Technological advances may accumulate at ever increasing rates for good reason. This may even provide a good explanation for the kind of idea-based growth that could allow for a more dematerialized future and sets growth the way economists typically describe it apart from material growth that ought to hit planetary limits. See Weitzman, "Recombinant Growth."

Page 19—**horse manure crisis**: The horse manure story has been told many times, perhaps most comprehensively by Eric Morris under the heading "From Horse Power to Horsepower," most prominently by Steven Levitt and Stephen Dubner in *SuperFreakonomics*, and most convincingly by Elizabeth Kolbert in a *New Yorker* review of the book's climate section ("Hosed"). Kolbert thankfully also clears up some of the misconceptions propagated by *SuperFreakonomics*.

This endnote thus far is fully taken from Wagner, *But Will the Planet Notice?* (which provides a further summary).

Page 20—**Nixon went on to sign:** Richard Nixon signed the National Environmental Policy Act of 1969 on January 1, 1970. It was the federal "Reorganization Plan No. 3" in July 1970 that led to the creation of the Environmental Protection Agency in December that year. Besides the acts listed, Nixon also signed the Federal Insecticide, Fungicide, and Rodenticide Act (1972), heavily amending the 1947 version that hadn't been very concerned with regulating pesticide use, the Noise Control Act (1972), and the Coastal Zone Management Act (1972). The official name of the "Clean Water Act" is the "Federal Water Pollution Control Act," which was amended in 1972.

Page 20—**local pollutants:** Axelrad et al., "Dose-Response Relationship," among many other studies, finds prenatal exposure to mercury correlates to loss of IQ (around .18 decrease in IQ per 1ppm increase in maternal hair mercury) by surveying the data from three previous studies. Brauer et al., "Air Pollution" finds a positive association between a child's exposure to soot, among other traffic-related air pollutants, and that child's risk of developing asthmatic and allergic symptoms as well as respiratory infections. Many early studies looked at what ingredients in smog are most inductive of eye irritation. See Altshuller, "Contribution of Chemical Species," and Haagen-Smit, "Los Angeles Smog." Long-term exposure to ozone in the troposphere, an essential ingredient of smog, has been linked to increased mortality rates. (Jerrett et al., "Ozone Exposure and Mortality"). The U.S. Safe Drinking Water Act authorizes the EPA to set standards for contaminants found in drinking water for good reason. See the U.S. EPA site for Drinking Water Contaminants (http://water .epa.gov/drink/contaminants/) for updated maximum permissible levels of contaminants, and more information of the health effects of different contaminants.

Page 20—**Niccolò Machiavelli:** The quote comes from Chapter VI of Machiavelli, *The Prince*, first distributed around 1515 and then published posthumously in 1532.

Page 21—**death:** Miller, *Coal Energy Systems*, and Rottenberg, *In the Kingdom of Coal*.

Page 21—**how people behave:** Refer back to "cognitive dissonance" on page 14. Moreover, collective action is particular difficult in the presence of persistent uncertainties. See Barrett, "Climate Treaties," which shows this point theoretically. Barrett and Dannenberg,

"Climate Negotiations," and Barrett and Dannenberg, "Sensitivity of Collective Action," confirm it experimentally.

Page 22—**greenhouse effect**: See the "Climate Science" entry on page 35 in chapter 2.

Page 22—**around 940** *billion* **tons**: See "tons of carbon dioxide" on page 93 in chapter 5.

Page 22—**2 ppm**: Rate of increase from CO_2Now (http://co2now.org /Current-CO2/CO2-Trend/acceleration-of-atmospheric-co2.html), calculated using source data from Keeling et al., Exchanges of atmospheric CO_2. In the past decade (2000 to 2010), greenhouse gas emissions have increased, on average, 2.2 percent a year, faster than during the three decades before 2000 (Summary for Policymakers of Working Group III of the *IPCC Fifth Assessment Report*). Carbon dioxide emissions from fossil fuel combustion and cement production alone have increased by 2.5 percent a year on average in the past decade (Friedlingstein et al., "Persistent Growth of CO_2 Emissions.")

Decadal averages may mask more recent changes in the trend. In 2012, for example, global carbon dioxide emissions rose by less than they had in an average year during the previous decade (and that despite the global recession of the late 2000s). Put differently, the increase in the increase in concentrations decreased in 2012. However, emissions still rose by around 1.4 percent (Olivier et al., "Trends in Global CO_2 Emissions"). Moreover, the hopeful trend did not continue in 2013, where emissions are projected to have risen 2.1 percent from 2012 (Le Quéré et al., "Global Carbon Budget 2013.")

Even if we slow the increase, stabilizing emissions (the inflow) will be far from enough. We need to stabilize (and eventually decrease) concentrations, the levels. Refer back to the earlier section on "The Bathtub Problem," beginning on page 15, and the "Bathtub" entry on page 30 in chapter 2.

Page 22—**billion or so high-emitters**: Chakravarty et al., "Sharing Global CO_2 Emission Reductions."

Page 22—**$500 billion per year**: See "World Energy Outlook 2014" for the latest country-specific numbers. The latest report puts the total at $548 billion in 2013, a $25 billion cut from the prior year. It also mentions that many countries are making moves to decrease their subsidies. Still, fossil fuel subsidies are over four times as high as subsidies for renewables. Meanwhile, global carbon dioxide emissions are at over 30 billion tons (World Resource Institute's Climate Analysis Indicators Tool). That averages out to subsidies of over $15 per ton of carbon dioxide. See Clements et al., *Energy Subsidy Reform*, for further estimates ($480 billion in 2011 for total energy subsidies) as well as "lessons and implications."

Contrast those subsidies with implicit carbon dioxide prices in certain countries due to other forms of regulation. Vivid Economics, "Implicit Price of Carbon," calculates the implicit carbon dioxide price in the electricity sectors in Australia, South Korea, China, Japan, the UK, and the United States. The price ranges from $0.50 per ton in South Korea, to $28.46 in the UK. The price in the United States is estimated to be around $5 per ton of carbon dioxide, roughly equal to total direct and indirect U.S. fossil fuel subsidies of around $3 per ton of carbon dioxide (OECD, "Fossil Fuel Subsidies," estimates that U.S. fossil fuel subsidies add up to around $16.3 billion in 2010).

Table 3 in Aldy and Pizer, "Comparability of Effort in International Climate Policy Architecture," presents carbon dioxide prices under various countries' energy and climate policies, ranging from the cap-and-trade program under the U.S. Regional Greenhouse Gas Initiative (RGGI) with a price of below $3 to German solar feed-in-tariffs estimated to be over $750 per ton of carbon dioxide abated.

Page 23—**stopped fuel subsidies**: "Nigeria Restores Fuel Subsidy to Quell Nationwide Protests."

Page 23—**Pigouvian taxes**: Pigou himself, it turns out, wrote about rabbits, not pollution: "incidental uncharged disservices are rendered to third parties when the game-preserving activities of one occupier involve the overrunning of a neighbouring occupier's land by rabbits— unless, indeed, the two occupiers stand in the relation of landlord and tenant, so that compensation is given in an adjustment of the rent." (Pigou, *The Economics of Welfare*.) But the principle is the same.

While Pigouvian taxes are, in fact, the efficient policy instrument, they also open up questions of redistribution. See, for example, Sterner, *Fuel Taxes*, which addresses questions of redistribution in the context of gasoline taxes.

Page 23—**possibly much more**: The precise numbers presented in the first table of the "Technical Update of the Social Cost of Carbon for Regulatory Impact Analysis under Executive Order 12866" for a ton of carbon dioxide emitted in 2015, using a 3 percent social discount rate, is $37. For 2020, the number is $43; for 2030, the number increases to $52. All figures are in inflation-adjusted 2007 dollars. The $37 figure would be much closer to $40 in today's dollars. The increase from $37 to $43 and $52 emphasizes the point that the damage caused by carbon dioxide is because of the concentration already in the atmosphere. The more is already there, the more marginal damage each additional unit causes.

This document referenced here represents the most recent update by the U.S. government, published November 1, 2013, and it marks a significant increase from figures published only three years prior. Back then, the central estimate of the social cost of carbon was $24 for a ton of carbon dioxide emitted in 2015. Table 1 of "Technical Update" summarizes the key factors that have led to the increase of the social cost between the 2010 and 2013 iterations. For DICE, those were "updated calibration of the carbon cycle model and explicit representation of sea level rise (SLR) and associated damages."

Also see Greenstone, Kopits, and Wolverton, "Developing a Social Cost of Carbon," for a detailed description of the original Interagency Working Group process that arrived at the 2010 estimate. In short, the U.S. government's social cost of carbon calculations are the result of a multiyear, multiagency review process, based on three well-established economic models. Among the most prominent such models is DICE, from Bill Nordhaus at Yale. For more on Nordhaus's model, see the "DICE" entry on page 36 in chapter 2 and the discussion around "$2 per ton" and "Nordhaus's preferred 'optimal' estimate" on page 57 in chapter 3.

For a detailed analysis of specific model shortcomings, see Kopp and Mignone, "Social Cost of Carbon Estimates." Van den Bergh and Botzen, "Lower Bound," argue for a social cost of at least $125 per ton of carbon dioxide. For a critique of integrated assessment models in general, see two prominent examples: Pindyck, "Climate Change Policy: What Do the Models Tell Us?" and Stern, "Structure of Economic Modeling." Pindyck's answer to the question posed in his title: "very little." Stern is similarly cautious about saying that economic models can tell us the full story. Both Pindyck and Stern, though, conclude that the Interagency Working Group's U.S. Social Cost of Carbon of around $40 per ton of carbon dioxide would be a good starting point. Stern declares it "far better than zero." Finally, for an argument tying social cost calculations to fat tails, the subject of chapter 3, see Weitzman, "Fat Tails and the Social Cost of Carbon."

Page 23—**35 cents per gallon:** The EPA estimates that each gallon of gasoline combusted produces on average 0.00892 metric tons of carbon dioxide. At $40 per ton, that is 35.68 cents per gallon. "Clean Energy: Calculations and References."

Page 24—**cap and trade:** Cap and trade was first introduced by Dales, *Pollution, Property, and Prices.* The United States used cap and trade to help remove chlorofluorocarbons in compliance with the Montreal Protocol, for getting lead out of gasoline, and, perhaps most prominently, for cutting sulfur dioxide from U.S. smokestacks in an effort to combat acid rain.

Page 24—**exact same result:** See Weitzman, "Prices vs. Quantities," for a theoretical argument around minimizing welfare losses under uncertainty. Newell and Pizer, "Regulating Stock Externalities under Uncertainty," extends the result by considering the case of a stock pollutant like carbon dioxide.

Page 24—**epic debates:** For a recent academic debate on taxes versus caps, see Keohane, "Cap and Trade, Rehabilitated" for the pro-cap argument, and Metcalf, "Designing a Carbon Tax," for the pro-tax argument. For a review of the debate, see Goulder and Schein, "Carbon Taxes vs. Cap and Trade."

Page 24—**cap and trade limits emissions:** See Keohane and Wagner, "Judge a Carbon Market."

Page 25—**lower compliance costs:** See, among others, Meng, "Estimating Cost of Climate Policy."

Page 25—**countervailing the force:** See Weitzman, "Negotiating a Uniform Carbon Price."

Page 27—**Electricity grid reform:** Harvard's Bill Hogan is a pioneer of this work. See, for example, Hogan, "Scarcity Pricing." For a good survey of grid reform more broadly, see Fox-Penner, *Smart Power*.

Page 27—**how cost-effective:** Karplus et al., "Vehicle Fuel Economy Standard," looks at fuel economy standards in comparison to and combination with emissions constraints. They estimate that the new U.S. CAFE standards will cost 6–14 times more than the fuel tax that would reach the same reduction in gas use. See also Fischer, Harrington, and Parry, "Automobile Fuel Economy Standards," for a good survey. See Jacobsen, "Evaluating U.S. Fuel Economy Standards," and Klier and Linn, "New-Vehicle Characteristics," for recent cost estimates of meeting CAFE targets. For a review of the effect of gasoline taxes, see Sterner, *Fuel Taxes*.

CHAPTER 2. 411

Page 31—**5 parts per million (ppm):** For the original data, see the Mauna Loa Observatory's data at http://www.esrl.noaa.gov/gmd/obop/mlo/. Also see "2 ppm" on page 22 in chapter 1.

Page 31—**700 ppm:** see "700 ppm" on page 14 in chapter 1.

Page 31—**400 ppm:** See "400 parts per million" on page 10 in chapter 1.

Page 32—**Some companies:** See Gunther, *Suck It Up*.

Page 32—**more expensive:** One flavor of this technology may have the potential to reverse the equation, at least in the narrow sense of those removing carbon: captured carbon dioxide could be piped underground to aid in pumping more oil. That comes under the term "enhanced oil recovery," and it turns captured carbon dioxide into a potentially valuable commodity. The irony—if that's the right term—is that it also leads to even more emissions.

Page 32—**Or maybe not**: The planet is experiencing unprecedented levels of technological progress, and for good reason (see, for example, Weitzman, "Recombinant Growth"). Morris, *Why the West Rules—for Now*, uses this fact to end his book on a debate that steers clear from his title question of whether the West, or China, will rule the future. Instead, Morris talks about the choice between "Singularity" and "Nightfall": how to avoid existential risks like climate change and navigate away from "Nightfall" and toward "Singularity."

Page 33—**linked to carbon dioxide**: See, for example, Shoemaker and Schrag, "Overvaluing Methane's Influence," and Solomon et al., "Atmospheric Composition." See also "decades of warming" and "excess carbon dioxide in the atmosphere" on pages 9 and 10, respectively, in chapter 1.

Page 34—**capping or taxing**: For the intricate yet often important differences between the two, see the debate on cap and trade versus taxation beginning on page 24 in chapter 1.

Page 34—**hit the mark**: See van Benthem, Gillingham, and Sweeney, "Learning-by-Doing."

Page 34—**fossil fuel subsidies**: See "$500 billion per year" on page 22 in chapter 1.

Page 35—**means subsidies**: For one of the best and most comprehensive arguments for a dual price-subsidy approach, see Acemoglu et al., "The Environment and Directed Technical Change."

Page 35—**discovered: 1824**: In the 1820s, Joseph Fourier calculated that, considering its distance from the sun, the earth should be much cooler than it is. Among other possible reasons for the extra heat, Fourier suggested that the atmosphere might somehow act as an insulator. (Fourier, "Remarques generals." The paper was republished three years later, with slight modifications: Fourier, "Les Temperatures.")

Page 35—**shown in a lab: 1859**: John Tyndall took Fourier's work a step further, when he began his lab experiments in January 1859 ("John Tyndall"). The seminal paper showing that gasses, including water vapor and carbon dioxide, could trap heat in the atmosphere was published in 1861 (Tyndall, "On the Absorption and Radiation of Heat").

Page 35—**quantified: 1896**: Svante Arrhenius first demonstrated the greenhouse effect and calculated climate sensitivity—what happens to temperatures as concentrations of carbon dioxide double—in 1896 (Arrhenius, "On the Influence of Carbonic Acid"). Arrhenius calculated climate sensitivity to be 5–6°C (9–11°F), larger than current consensus estimates of between 1.5 and 4.5°C (2.7 and 8.1°F), established in the 1970s (Charney et al., "Carbon Dioxide and Climate"). See the extensive discussion on climate sensitivity in chapter 3 for more.

Page 35—**Climate Sensitivity**: "Climate sensitivity" or "equilibrium climate sensitivity" is widely defined as the global average surface equilibrium temperature warming from a doubling of atmospheric carbon dioxide concentrations. It is inherently a long-run estimate, how temperatures will react over many decades or centuries, in "equilibrium." In geological times, this still counts as "fast." See "decades to centuries" on page 10 in chapter 1 on the distinction between "fast equilibrium"—what is captured by most commonly used climate sensitivity parameters—and so-called earth system sensitivity, which could be over double the prevailing climate sensitivity estimate. Climate sensitivity ranges are typically pieced together from various estimates: actual temperatures measured by instruments over the past 150 or so years; paleoclimatic evidence from glacial and other developments over the past millions of years; carefully calibrated climate models; and a host of other means like evidence from volcanic eruptions or simple expert elicitation (asking climate scientists about their best guesses). For a comprehensive review, see Knutti and Hegerl, "Equilibrium Sensitivity."

See chapter 3 on "Fat Tails" for the history of climate sensitivity calculations and the profound implications.

Page 36—**DICE**: Bill Nordhaus first introduced DICE in 1991. A later derivation, RICE, includes regional differences. See, for example, Nordhaus, "To Slow or Not to Slow," and, most prominently, Nordhaus, "Optimal Transition Path." For the most comprehensive description at the time, see Nordhaus, "Optimal Greenhouse Gas Reductions." The latest and most comprehensive description of his work is in Nordhaus, *Climate Casino*. For a later update, see Nordhaus, "Estimates of the Social Cost of Carbon," which arrives at a price of $18.6 per ton of carbon dioxide emitted in 2015 (in 2005 dollars). For further in-depth discussion see "possibly much more" on page 23 in chapter 1 and the discussion on page 57 in chapter 3, around "$2 per ton" and "Nordhaus's preferred 'optimal' estimate."

Page 35—**Free Drivers**: Our definition and subsequent use of the term solely focuses on the context of geoengineering. Others have used the term in the context of energy-efficiency improvements, referring to a type of network effect, and a highly positive one at that: those outside a particular energy-efficiency program may adopt the more efficient technology because they feel compelled to do so by participants in the program. See, for example, Gillingham, Newell, and Palmer, "Energy Efficiency Economics."

Page 38—**Mount Tambora**: Stothers, "The Great Tambora Eruption," estimates a mean temperature decrease in the Northern Hemisphere of 0.4–0.7°C (0.7–1.3°F). For a more detailed description of the eruption and its diverse range of consequences, see Klingman and Klingman, *The Year without a Summer*, and Stommel, *Volcano Weather*.

Page 39—**heart of the global problem**: For one of the first and the most widely cited explorations of the subject, see Hardin, "Tragedy of the Commons."

Page 40—**centuries and millennia**: See "excess carbon dioxide in the atmosphere" on page 10 in chapter 1.

Page 41—**over 10 meters**: See "Melting of Greenland" on page 56 in chapter 3.

Page 41—**10 percent**: See, for example, McGranahan, Balk, and Anderson, "The Rising Tide"; Anthoff et al., "Global and Regional Exposure"; Rowley et al., "Risk of Rising Sea Level."

Page 41—**one seeming exception**: Jensen and Miller, "Giffen Behavior and Subsistence Consumption."

Page 42—**Ocean Acidification**: The report "Economics of Ocean Acidification," from an International Atomic Energy Agency international workshop in 2012, provides an overview of the economic impacts of ocean acidification, while IGBP, IOC, SCOR, "Ocean Acidification Summary for Policymakers," discusses the science of the phenomenon. For a good summary of both, see "Acid Test." For more on the marine die-off 56 million years ago, see Thomas, "Biogeography of the Late Paleocene." There was no associated mass die-off in the terrestrial biosphere. Cui et al., "Slow Release of Fossil Carbon," shows how the peak rate of carbon dioxide release into the atmosphere around 56 million years ago was much lower than today.

Page 42—**alkalinity addition**: See Harvey, "Mitigating the Atmospheric CO_2 Increase," for a comprehensive discussion of directly adding limestone powder to oceans. See Rau, "CO_2 Mitigation," for a different method that involves first capturing carbon dioxide on land before releasing the alkaline solution into the ocean. Royal Society, "Geoengineering the Climate," includes a brief discussion and puts it in the larger context.

Page 44—**60 percent of global emissions**: Calculated for 2010, using the World Resource Institute's Climate Analysis Indicators Tool.

Page 44—**Books:** Sunstein's "Of Montreal and Kyoto," and the later adaptation in Sunstein, *Worst-Case Scenarios*, provide a comparative history and analysis of the Montreal and Kyoto Protocols. He suggests a few reasons for why the former worked so well while the latter has at best led to small steps in the right direction. In particular, Sunstein makes a strong case that success of the one, and the failure of the other, had a lot to do with domestic benefit-cost analysis in the United States. For a terrific insider's view on the making of the Montreal Protocol, see Benedick, *Ozone Diplomacy*. Barrett, *Environment and Statecraft*, uses, in part, the success of the Montreal Protocol to develop a theory on international environmental treaties, and what makes them work, or, in most cases, fail.

CHAPTER 3. FAT TAILS

Page 48—**more likely than not:** The IPCC attempts to assign plain-English terms to its consensus assessments: "more likely than not" corresponds to a likelihood of greater than 50 percent; "likely" corresponds to greater than 66 percent (not two-thirds; i.e. 67 percent); "very likely" corresponds to greater than 90 percent; "extremely likely" corresponds to greater than 95 percent. These terms were used to describe the likelihood of man-made global warming in, respectively, the IPCC's Second, Third, Fourth, and Fifth Assessment Reports. According to Engber, "You're Getting Warmer," an early draft of the Fourth Assessment Report called for the highest category, "virtually certain," which corresponds to greater than 99 percent probability, before settling at "very likely." Engber discusses the history and implications of the IPCC's probability assessments. For the latest formal guidance document, see Mastrandrea et al., "IPCC AR5 Guidance Note." For more on the history and the scientific underpinnings, see Giles, "Scientific Uncertainty." For a survey of how these probabilistic statements are perceived (and often misconstrued), see Budescu et al., "Interpretation of IPCC."

Page 48—**decade 'without warming':** For more on the warming "pause" or "hiatus" of recent years, see "warmest in human history" on page 9 in chapter 1.

Page 48—**back to the 1800s:** See "Climate Science" on page 35 in chapter 2. For more on the history, and the future, see Roston, *The Carbon Age*.

Page 49—**Wally Broecker:** Broecker, "Climatic Change."

Page 49—**climate sensitivity:** See "Climate Sensitivity" on page 35 in chapter 2.

Page 49—**well-established facts**: Stocker, "Closing Door," and Matthews et al., "Proportionality of Global Warming," are among the latest to discuss the proportional relationship between total warming and cumulative emissions.

Page 50—**400 ppm**: This is the concentration of carbon dioxide. Counting other greenhouse gases (without aerosols), concentrations are between 440 and 480 ppm of carbon dioxide–equivalent greenhouse gases, depending on the source. See "400 parts per million" on page 10 and also "2 ppm" on page 22 in chapter 1.

Page 50—**700 ppm**: See "700 ppm" on page 14 in chapter 1.

Page 50—**Ad Hoc Study Group**: Charney et al., "Carbon Dioxide and Climate."

Page 50—**academic genius**: Gavin Schmidt tells the story on Real Climate.org, an excellent repository of the latest on climate change science (Schmidt and Rahmstorf, "11°C Warming").

Page 50—**the "likely" range**: By 1990 the IPCC range was still 1.5–4.5°C (2.7–8.1°F). Ditto by 1995 and 2001. By 2007, the range narrowed somewhat, though in the wrong direction. It seemed that 1.5°C (2.7°F) was no longer in the cards. The new "likely" range was 2–4.5°C (3.6–8.1°F). By 2013, the most recent IPCC Assessment report, the range widened again right back to where it's been all along: 1.5–4.5°C (2.7–8.1°F). For the relevant sections of the reports, see Working Group 1, chapter 5, of the *IPCC First Assessment Report*, Section B: Climate Modelling, Climate Prediction and Model Validation, of the *IPCC Climate Change 1992* Supplementary Report, Working Group I of the *IPCC Second Assessment Report*, Working Group I of the *IPCC Third Assessment Report*, Working Group I of the *IPCC Fourth Assessment Report*, and Working Group I of the *IPCC Fifth Assessment Report*.

It is true that the confidence in the range has increased markedly over time. Specifically, "confidence today is much higher as a result of high quality and longer observational records with a clearer anthropogenic signal, better process understanding, more and better understood evidence from paleoclimate reconstructions, and better climate models with higher resolution that capture many more processes more realistically" (Working Group I of the *IPCC Fifth Assessment Report*, TFE.6; also see Box 12.2). Still, the IPCC chose to call the range "likely" (>66 percent confidence) rather than opt for a more certain assessment such as "very likely" (>90 percent).

Things may even be worse than before for another reason. In 1990, the IPCC ventured a "best guess" of 2.5°C (4.5°F) within the wider range. By 2007, the "most likely" quantity was 3°C (5.4°F). Not certainty, not even an actual "mean" or "median" in statistical terms, but at least a single number—albeit a high one—around which to rally. By 2013, the IPCC issued no verdict as to which quantity would be most likely. That's a step back in sureness. The IPCC did add other caveats, notably a less than 5 percent probability of climate sensitivity being below 1°C and a less than 10 percent probability of above 6°C. See "clearly more room" on page 51 as well as "more likely than not" on page 48 for a definition of the "likely" range itself.

Page 51—**defines "likely"**: See "more likely than not" on page 48 as well as prior note.

Page 51—**clearly more room**: The IPCC's latest assessment report goes into a bit more detail: it describes anything below 1°C as "extremely unlikely" (0–5 percent) and anything above 6°C as "very unlikely" (0–10 percent) (Summary for Policymakers of Working Group I in the *IPCC Fifth Assessment Report*). The second row of this table translates the IPCC's statements into actual probabilities for different climate sensitivities:

Climate sensitivity	*<0°C*	*<1°C*	*<1.5°C*	*<>2.6°C*	*>3°C*	*>4.5°C*	*>6°C*
IPCC (2013)	No data	0–5%	("likely" between 1.5–4.5°C)				0–10%
Our calibration	0%	1.7%	11%	50% (78% probability of between 1.5–4.5°C)	37%	11%	3.1%

We calibrated a log-normal distribution by calculating an 11 percent probability of being greater than 4.5°C and an 11 percent probability of being below 1.5°C. Doing so interprets the IPCC's numbers as conservatively as possible. The IPCC, for example, states that any figure above 6°C would be "very unlikely." That implies a 0–10 percent range—5 percent, if we take a point estimate. However, if the IPCC authors wanted to say that it was, in fact, only 5 percent, they could have chosen to say "extremely unlikely." By saying "very unlikely," they, in effect, may have intended to ascribe a probability of between 5 and 10 percent—7.5 percent as the point estimate. Either way, our calibration arrives at a probability estimate of slightly over 3 percent for the chance of climate sensitivity being greater than 6°C, a "conservative" estimate for the purposes of our exercise that remains much below 7.5 percent.

Our interpretation of the "likely" range uses a similar logic: The IPCC definition of "likely" is between 66 and 100 percent. However, if the authors wanted to convey that the probability of being in the 1.5–4.5°C range was higher than 90 percent, they could have chosen to call the range "very likely." (In fact, "very likely" does have a firm definition in the guidance document for IPCC authors, while "extremely likely" is an additional term added by the authors involved in Working Group I of the *IPCC Fifth Assessment Report* (see Working Group I's Summary for Policymakers). For comparison, see Mastrandrea et al., "IPCC AR5 Guidance Note." Instead, the IPCC authors opted for the looser interpretation of "likely," which leads us to believe that the true likelihood may not be between 66 and 100 percent but between 66 and 90 percent. We split the difference and use 78 percent with 11 percent probability of being below the likely range and 11 percent probability of being above the likely range.

Our median estimate is 2.6°C (3.9°F). The most commonly cited mean climate sensitivity, 3°C (5.8°F), is therefore closer to the two-thirds mark in our calibration: we assume a 63 percent probability of climate sensitivity being below 3°C (5.8°F) and, conversely, a 37 percent probability of climate sensitivity being above 3°C (5.8°F). The latter is the probability mentioned in the table. The remaining estimates are in the bottom row of the table above.

Recent papers in *Science* and *Nature* have made the argument that climate sensitivity is much more likely above 3°C than below it. Fasullo and Trenberth, "A Less Cloudy Future," finds that models with lower climate sensitivities do not fully take into account albedo from changing cloud cover. Sherwood, Bony, and Dufresne, "Spread in Model," takes this idea further, and posits that accurate accounting of the cloud mixing processes suggests a climate sensitivity greater than 3°C (5.8°F).

Page 51—**bottle of Champagne:** A lot of analysis has gone into discovering the delicate science behind a good bottle of champagne. The average 750 ml bottle contains 9 grams of carbon dioxide, and emits about five liters once opened (Liger-Belair, Polidori, and Jeandet, "Science of Champagne Bubbles"). Not to mention the 200,000 metric tons released each year transporting the bubbly around the world (Alderman, "A Greener Champagne Bottle"). One way to minimize the loss of dissolved carbon dioxide in your champagne is to pour it like you would a beer—down the side of the glass instead of right to the bottom. It may not be quite as sophisticated, but scientists assure it will make for a better taste (Liger-Belair et al., "Losses of Dissolved CO_2"). Ironically, the celebratory champagne might not taste as nicely as it would if we had no cause for celebration at all. Wine quality in the Champagne region of France is, in fact, expected to increase based on projected climate change (Jones et al., "Global Wine Quality").

Page 52—**fat tail**: The technical definition of a "fat tail" is a distribution that approaches zero polynomially or slower. Conversely, the technical definition of a "thin tail" is a distribution that approaches zero exponentially or faster. A log-normal distribution, which we use, is in between thin and fat tailed. Some definitions call it "heavy tailed": no longer thin, but not yet fat either. Log-normal distributions approach zero faster than polynomially but slower than exponentially. All that means that our calibration is more conservative than the IPCC's own numbers, as it points to a chance of slightly above 3 percent of climate sensitivity being above 6°C. That compares to the IPCC's stated "very unlikely" range of anywhere between 0 and 10 percent. See "clearly more room" on page 51.

Page 53—**scientific papers**: The conceptual starting point for calculations leading up to this table is Weitzman, "Modeling and Interpreting the Economics." A further elaboration is found in Weitzman, "Fat-Tailed Uncertainty." A version resembling this table, with calculations for three different probability distributions but based on the *IPCC Fourth Assessment Report*, appeared in Weitzman, "GHG Targets as Insurance." The version here is based on fitting a log-normal probability distribution to climate sensitivity, as described in "clearly more room" on page 51:

CO_2e concentration (ppm)	400	450	500	550	600	650	700	750	800
Temperature increase for mean climate sensitivity (=3°C)	1.5°C	2.1°C	2.5°C	2.9°C	3.3°C	3.6°C	4.0°C	4.3°C	4.5°C
Temperature increase for median climate sensitivity (=2.6°C)	1.3°C	1.8°C	2.2°C	2.5°C	2.7°C	3.2°C	3.4°C	3.7°C	3.9°C
Chance of >6°C, given 2–4.5°C "likely" range (with 70% probability)	0.03%	0.3%	1.3%	3.3%	6.3%	10.2%	14.4%	19.2%	23.9%

CO_2e concentration (ppm)	400	450	500	550	600	650	700	750	800
Chance of >6°C, given 1.5–4.5°C "likely" range (with 70% probability)	0.2%	1.1%	2.8%	5.2%	8.1%	11.3%	14.6%	18.0%	21.3%
Chance of >6°C, given 1.5–4.5°C "likely" range (with 78% probability)	0.04%	0.3%	1.2%	2.7%	4.9%	7.6%	10.6%	13.9%	17.3%

Row 1 has concentrations of ultimate carbon dioxide–equivalent (CO_2e) concentrations. Row 2 shows final temperature increases based on these concentrations and an assumed climate sensitivity of 3°C, the median figure found from a calibration to the 2007 *IPCC Fourth Assessment Report* "likely" range of 2–4.5°C. Row 3 shows the temperature increase for an assumed climate sensitivity of 2.6°C, the median figure found for our "best" log-normal calibration to the 2013 *IPCC Fifth Assessment Report* "likely" range of 1.5–4.5°C. The latter is also what we present in the main text. Note that this single "most likely" temperature increase is below the average or expected figure. That's because the distribution fitted around the IPCC "likely" range for climate sensitivity is assumed to be an asymmetric log-normal distribution, which cuts off at zero but has a long upward tail. Despite the uncertainties, no one would seriously argue that climate sensitivity should have a negative realization. The next three rows present various assumptions around the IPCC "likely" range: Row 4 assumes the old climate sensitivity range of 2–4.5°C, the recent consensus before the release of the *IPCC Fifth Assessment Report* in 2013. Row 5 widens the "likely" range to extend to 1.5°C on the lower end. Both rows assume a 70 percent probability that climate sensitivity is within the "likely" range, rounding up the IPCC's 66 percent number for its definition of "likely." The final row then splits the difference between 66 percent ("likely") and 90 percent ("very likely") to put a probability of 78 percent of being between 1.5 and 4.5°C and a probability of 11 percent of climate sensitivity being above 4.5°C. This works out to a probability of 3 percent that climate sensitivity is above 6°C, which is conservatively low by almost any standards. The final row is what we present in the main text, with numbers further rounded for simplicity.

Page 55—**heavy**: See "fat tail" on page 52.

Page 55—**the median**: Note that we are using the median climate sensitivity of 2.6°C here to calculate temperature increase. Using the more conventional mean climate sensitivity of 3°C would translate into a (mean) temperature increase of 4.0°C based on concentrations reaching 700 ppm (rather than the median figure of 3.4°C presented in the main text).

Page 55—**Black Swan**: Taleb, *Black Swan*.

Page 55—**unknown unknowns**: Donald Rumsfeld popularized the term in the context of the U.S. invasion of Iraq, drawing the analogy at more than one occasion. The first mention was at a Pentagon news conference on February 12, 2002: "Reports that say that something hasn't happened are always interesting to me, because as we know, there are known knowns; there are things we know we know. We also know there are known unknowns; that is to say we know there are some things we do not know. But there are also unknown unknowns—the ones we don't know we don't know. And if one looks throughout the history of our country and other free countries, it is the latter category that tend to be the difficult ones" (Morris, "Certainty of Donald Rumsfeld"). Rumsfeld later echoed the sentiment at least once more, at a NATO press conference on June 6, 2002 (Rumsfeld, "Press Conference").

Economists typically credit Chicago economist Frank Knight for coming up with the idea (Knight, "Risk, Uncertainty, and Profit"). He made the technical distinction between "risk" and "uncertainty." (Note that Knightian "risk" is different from what the average person—including the average scientist—calls "risk." Any layman's "existential risk," including the way we use it in the text, is much closer to Knightian "uncertainty" than Knightian "risk.") Richard Zeckhauser has added a third category: "ignorance." Risk deals with known distributions. Uncertainty is not knowing which distribution to pick. Ignorance is when it's unclear there even is a distribution. See Zeckhauser, "Unknown and Unknowable," and a subsequent reaction: Summers, "Comments."

Page 56—**bad global warming feedbacks**: Walter et al., "Methane Bubbling," attempts to measure methane emissions from thaw lakes in Siberia, estimating that methane emissions from northern wetlands is 10–60 percent higher than previously thought. They find that the largest portion of the methane released comes from the thawing permafrost around lake edges. This process is thought to be a critical one in previous times of climatic change. "Climate Science: Vast Costs of Arctic Change" estimates the total cost to society of methane released from thawing Siberian permafrost to be on the scale of $60 trillion.

Page 56—**Melting of Greenland**: The Greenland ice sheet has a sea level equivalence of 7.36 m (24 feet), and full melting of the Antarctic ice sheet would mean a 58.3 m (191 feet) sea-level rise (see chapter 4, "Observations: Cryosphere," in Working Group I of the *IPCC Fifth Assessment Report*). Full melting of the West Antarctic ice sheet by itself would lead to about 3.3 meters (11 feet) of sea-level rise (Bamber et al., "Potential Sea-Level Rise"). See "centuries of sea-level rise" on page 9 in chapter 1 for more on the irreversibility of it melting.

The *IPCC Fifth Assessment Report* found that the Greenland ice sheet has contributed on average 0.59 mm per year to sea-level rise from 2002 to 2011, while the Antarctica contribution is likely 0.4 mm per year for the same period. Both of these contribution rates have more than quadrupled from the average for 1992 to 2001. The observed global mean sea-level rise for the 1993 to 2010 period was 3.2 mm per year. The IPCC's estimate for total sea-level rise under the worst-case scenario is 0.53 to 0.97 m (1.7 to 3.2 feet). The only situation they believe could increase sea level by 2100 significantly above this likely range would be if marine-based sections of the Antarctic ice sheet collapsed (see chapter 13, "Sea Level Change" in Working Group I of the *IPCC Fifth Assessment Report*).

Page 57—**DICE model**: See the "DICE" entry on page 36 in chapter 2.

Page 57—**$2 per ton**: Nordhaus derives a number of $5 per ton of *carbon* for 1990 to 1999 (Nordhaus, "Optimal Transition Path"). We convert this figure to dollars per ton of *carbon dioxide* and into 2014 dollars using the GDP deflator to arrive at $2 per ton of carbon dioxide.

Page 57—**Nordhaus's preferred "optimal" estimate**: Nordhaus, "Estimates of the Social Cost of Carbon," presents a price of $18.6 per ton of carbon dioxide emitted in 2015 (in 2005 dollars). Converted into 2014 dollars, the figure is around $20 per ton. The paper presents both what Nordhaus considers the "optimal" path and various other scenarios, including one to keep global average temperature increases below 2°C. Note that this $20 estimate is significantly higher than his "optimal" path derived only four years prior. Then the optimal figure for 2015 was $12 (Nordhaus, "Economic Aspects"). Note also that the $20 is lower than both Nordhaus's set of "illustrative carbon prices needed for a 2½°C temperature limit" (figure 33 in

Nordhaus, *Climate Casino*) and the "central" estimate presented in the first table of the "Technical Update of the Social Cost of Carbon for Regulatory Impact Analysis under Executive Order 12866" for a ton of carbon dioxide emitted in 2015. The former is $25 for a ton emitted in 2015. The latter is close to $40 per ton, using an average of three models and a discount rate of 3 percent. Nordhaus's own preferred discount rate is around 4.2 percent. He shows how the difference in discount rates explains most of the difference between the $40 and his own $20 estimate.

Page 58—**around $40**: See "possibly much more" on page 23 in chapter 1.

Page 59—**more equable and better climates**: The full quote: "By the increasing percentage of [carbon dioxide] in the atmosphere, we may hope to enjoy ages with more equable and better climates, especially with regards to the colder regions of the earth, ages when the earth will bring forth much more abundant crops than at present, for the benefit of rapidly propagating mankind" (Arrhenius, *Worlds in the Making*, 63).

The general spirit of the importance of and opportunities in adapting to warmer climates is best represented in Kahn's *Climatopolis*. There are surely costs, Kahn argues, but coping mechanisms create their own opportunities, especially for highly efficient cities.

Page 59—**playing catch-up**: See "$2 per ton" and "Nordhaus's preferred "optimal" estimate" on page 57 for a discussion of the evolution of Nordhaus's DICE estimates. See "possibly much more" on page 23 in chapter 1 for a discussion of the U.S. government's figures.

Page 60—**started by one person**: Bill Nordhaus created DICE. Richard Tol developed FUND, which is now largely maintained by David Anthoff: http://www.fund-model.org/. Chris Hope was the driving force behind PAGE: http://climatecolab.org/resources/-/wiki/Main /PAGE.

Page 60—**massive data operations**: The underlying global circulation models used by climate scientists and feeding into the IPCC reports are indeed computationally complex. However, integrated assessment models then rely on much-simplified output, in DICE's case on the Model for the Assessment of Greenhouse Gas Induced Climate Change (MAGICC), which in turn is a much-simplified version of underlying climate models. DICE itself is freely available on Bill Nordhaus's website and even runs in Excel: http://www.econ.yale .edu/~nordhaus/.

Page 61—**Lots are missing**: See, for example, Howard, "Omitted Damages." Van den Bergh and Botzen, "Lower Bound," similarly present climate change effects that are inadequately captured by models like DICE. Most of these effects would increase estimates of the social cost of carbon. Some may also decrease it. (See "possibly much more" on page 23 in chapter 1.)

Page 61—**quadratic extrapolations**: DICE uses an inverse quadratic loss function linked to temperature (T), where loss is defined as equal to $1/[1+a\,T+b\,T^2]$.

Page 62—**as far out as 6°C**: In fact, Nordhaus, *Climate Casino*, cuts off the graph at 5°C (9°F), implying that any damages due to temperature changes beyond that level are much too uncertain (or perhaps rare) to contemplate.

Page 63—**damages affect output growth** *rates*: See Pindyck, "Climate Change Policy," and Heal and Park, "Feeling the Heat," who link temperature to labor productivity via human physiology. They find high temperatures decrease productivity in already hot—and often poor—countries, while higher-than-average temperatures increase productivity by similar amounts in cool—and typically rich—countries.

Moyer et al., "Climate Impacts on Economic Growth," similarly show large impacts on the Social Cost of Carbon estimate of changing the climate impact from output levels to productivity: "even a modest impact of this type increases SCC estimates by many orders of magnitude."

Page 65—**damages are additive**: See Weitzman, "Damages Function." For a complementary take—focusing on the idea of "relative prices"—see Sterner and Persson, "Even Sterner Review." The fundamental distinction between multiplicative versus additive damage functions rests on questions of substitutability. The (implicit) assumption for multiplicative damages is unit substitutability between economic sectors and environmental amenities within the utility function. Additive damages assume less (to no) substitutability across these sectors in the utility function.

Page 68—**likely underestimate**: See "possibly much more" on page 23 in chapter 1.

Page 68—**discount whichever number**: Many a book and article has been written on the topic. Gollier, *Pricing the Planet's Future*, ranks among the best general introductions.

Page 68—**is worth more**: In fact, having one dollar today is typically worth a lot more than having it tomorrow. Whereas the same one-day difference a hundred years from now is barely noticeable. From today's perspective, a hundred years plus one day is pretty much the same as a hundred years. Quite naturally, humans tend to discount the first day more heavily than that one day a hundred years from now. The technical term for this particular phenomenon is "hyperbolic discounting," most prominently introduced to economics in Laibson, "Golden Eggs."

Page 69—**2 percent a year**: "10-Year Treasury Inflation-Indexed Security, Constant Maturity."

Page 69—**criticized for that low choice**: Weitzman, "A Review," argues that "the *Stern Review* may well be right for the wrong reasons," the low discount rate being one of the wrong reasons.

Page 69—**decline over time**: See Weitzman, "Gamma Discounting," for the declining discount rate numbers mentioned in the text. For a later consensus view around the logic behind declining discount rates, not necessarily the specific numbers, see Arrow et al., "Determining Benefits and Costs," and Cropper et al., "Declining Discount Rates." France and the United Kingdom, for example, use declining discount rates, but they do not agree on the exact rate: France's starts at 4 percent and declines to a bit over 2 percent for numbers 300 years in the future; the UK's starts at 3.5 percent and declines to 1 percent after 300 years.

For an application that reconciles some important technical differences in the application of the basic logic, see Gollier and Weitzman, "Distant Future." It concludes that "The long run discount rate declines over time toward its lowest possible value."

To see the reason behind this declining rate, consider the following thought experiment: Assume we don't know whether the true discount rate for damages a hundred years from now should be 1 percent or 7 percent. The former is on the lower end of rates for U.S. Treasury bills, which come as close to a risk-free investment as possible. The latter comes from the obscure but all-important "Circular A-94," which the powerful U.S. government's Office of Management and Budget suggests as the base-case analysis for all government investment and regulatory decisions (OMB, "Circular No. A-94 Revised"). Note that the 7 percent rate is not a risk-free rate in any sense of that word. In fact, it deliberately deals with risky investment decisions. The big question there is whether the rate for riskier investments ought to go up or down, something discussed in more detail in the text itself.

OMB's other, and more appropriate, rate for something as far out as a hundred years is 3 percent. That's also the government's base case for the $40 social cost of carbon. But for argument's sake, let's use the 7 percent figure for now. It's certainly an upper bound of sorts. Hardly anyone would sensibly argue for a higher discount rate. There's a good reason why: $100 a hundred years from now, discounted at 7 percent is worth 9 cents today. Invest less than a dime today at a rate of return of 7 percent, and expect to get $100 a century hence. For an investor, that's not bad. For discounting climate damages a hundred years out, it makes them almost worthless today. That, of course, is the exact line of reasoning some use to argue why climate damages don't matter all that much. Why worry about costs of global warming in a century, if all it takes is setting aside relatively little money today to cover the damages? All that holds true at 7 percent. But $100 a hundred years from now, discounted at 1 percent, is worth $37 today. That's quite a bit more.

Let's split the difference and use a 4 percent discount rate, halfway between 1 and 7 percent: now $100 a hundred years from now is worth $1.8 today. That's much, much closer to 9 cents from the 7 percent discount rate than the $37 from a 1 percent rate. But that's only one way of "splitting the difference." What if we just didn't know whether the rate should be 1 percent or 7 percent?

Put the chance of either rate at 50–50. That's a 50 percent probability that the correct number should be 9 cents, and a 50 percent probability that it should be $37. On average, that's roughly $18. That average of the discounted numbers is much higher than the number using the average discount rate of 4 percent = (7 percent + 1 percent)/2. In fact, in our example, the difference is a factor of ten: $18 versus $1.80. And the difference increases the further out you go.

Lastly, for a different argument for declining discount rates, see Heal and Millner, "Agreeing to Disagree." They suggest that the choice of discount rate is an "ethical primitive" to arrive at the same conclusion of declining discount rates.

Page 70—**Black-Litterman Global Asset Allocation Model**: Black and Litterman, "Global Portfolio Optimization."

Page 71—**stark implications**: "If climate risk dominates economic growth risk because there are enough potential scenarios with catastrophic damages, then the appropriate discount rate for emissions investments is lower tha[n] the risk-free rate and the current price of carbon dioxide emissions should be higher. In those scenarios, the 'beta' of climate risk is a large negative number and emissions mitigation investments provide insurance benefits. If, on the other hand, growth risk is always dominant because catastrophic damages are essentially impossible and minor climate damages are more likely to occur when growth is strong, times are good, and marginal utility is low, then the 'beta' of climate risk is positive, the discount rate should be higher than the risk-free rate, and the price of carbon dioxide emissions should be lower" (Litterman, "Right Price"). For earlier use of "*beta*" in the climate context, see Sandsmark and Vennemo, "Portfolio Approach," for an argument for a negative *beta* of mitigation investments. Gollier, "Evaluation of Long-Dated Investments," makes the case for a positive *beta*.

Page 72—**equity premium puzzle**: For an overview, see Mehra, "Equity Premium Puzzle."

Page 72—**reverses the equity premium puzzle**: For a technical exploration of this argument, see Weitzman, "Subjective Expectations," or Barro, "Rare Disasters." See Mehra, "Equity Premium Puzzle," for alternative explanations of the equity premium puzzle and a survey of the ongoing debate.

Page 72—**"black" days**: *Black Monday* on October 19, 1987, saw the Dow drop 22 percent; *Black Tuesday* on October 29, 1929, marked the beginning of the Great Depression; *Black Wednesday* on September 16, 1992, earned George Soros a billion pounds betting against the Bank of England; *Black Thursday* on October 24, 1929, saw Wall Street lose over 10 percent almost at the opening bell (recall *Black Tuesday* just above for what happened next); *Black Friday* on September 24, 1869, saw markets crash after a failed attempt to corner the gold market. None of it should be confused with the *Black Week* beginning Monday, October 6, 2008, when the Dow fell 18 percent by Friday. The front-page articles of the *Wall Street Journal* after each of these events make for interesting, contemporaneous reading: For the reaction to what we now know as "Black Monday," see Metz et al., "Crash of '87." After Black Tuesday and Thursday, the two days that are seen as the beginning of the Great Depression, the WSJ seems surprisingly nonchalant. "Pressure Continues: Stocks Sink Lower under Record Volume of Liquidation," (published October 30, 1929) recognizes the huge drop in stock prices from the day before, but also states that "industrial activity is on a large scale and sound basis with no real indications of a depression in prospect," and projects that "after the initial shock has worn off the decline will prove beneficial in many ways by releasing funds from market to industry." "Demoralized Trading: Stocks Break on Record Volume—Banking Support Starts Rally" (published October 25, 1929), somewhat more awed by the situation, opens with the statement, "Yesterday's market was in many respects the most extraordinary in the history of the Stock Exchange." However, the article also ends with a projection that the market would turn around soon. Zweig, "What History Tells Us," looks at the preceding "Black Week" in comparison to and in the context of the "Great Crash." For the front-page article of London's *Financial Times* on the day after Black Wednesday, see Stephens, "Major Puts ERM Membership on Indefinite Hold."

Page 73—**drawing the link**: Litterman, "Right Price."

Page 74—**won't happen tomorrow**: We have a much better idea of warming in the short and medium term: For the next two decades (2016 to 2035), the Summary for Policymakers in Working Group I of the *IPCC Fifth Assessment Report* finds, with "medium confidence," a "likely" additional warming of between 0.3 and 0.7°C (0.5 and 1.3°F) relative to the past two decades (1986–2005).

For the final two decades of this century, the predictions diverge dramatically. Depending on which scenario one chooses, average global warming relative to the past two decades could be anywhere from 0.3–1.7°C, to 2.6–4.8°C (0.5–3.1°F, to 4.7–8.6°F), an enormous range with dramatically different consequences. And those are just the likely ranges. See, for example, "0.3 to 1 meters" on page 5 of chapter 1 for the implications for sea-level rise.

Note that all these estimates, including for sea-level rise, are relative to the two decades ending in 2005. 4.8°C (8.6°F) of additional warming relative to "today" would mean total warming of 5.5°C (9.9°F) from preindustrial levels.

Page 74—**the longer it will take**: See Roe and Bauman, "Climate Sensitivity," for this point. They employ a standard willingness-to-pay framework to conclude that fat tails may not be that costly. (For a contrasting conclusion by the same [lead] author, see Roe, "Costing the Earth.")

Page 77—**European Union**: See Ellerman, Convery, and de Perthuis's *Pricing Carbon* for an early yet comprehensive survey of the EU's emissions trading system.

Page 77—**sole exception is Sweden**: See Hammar, Sterner, and Åkerfeldt, "Sweden's CO_2 Tax," and Johansson, "Economic Instruments in Practice."

Page 77—**decision criterion**: For a recent elaboration on the point around alternative decision criteria, see Heal and Millner, "Uncertainty and Decision." Also see Millner, Dietz, and Heal, "Scientific Ambiguity and Climate Policy."

Page 78—**ethical component**: For a climate scientist making the strong moral case, see Roe, "Costing the Earth." For a moral philosopher making the strong case for economists to engage on the moral dimension, see Sandel, "Market Reasoning as Moral Reasoning."

CHAPTER 4. WILLFUL BLINDNESS

Page 80—**U.S. Supreme Court**: *Global-Tech Appliances, Inc., et al. v. SEB S.A.*

Page 81—**colloquial interpretation**: For a popular take on "willful blindness"—with some brief cameos by climate change—see Heffernan, *Willful Blindness*.

Page 81—**tax or cap carbon**: See our discussion of taxes versus cap and trade in chapter 1.

Page 81—**close to zero**: See "$500 billion per year" on page 22 in chapter 1.

Page 81—**around $40**: See "possibly much more" on page 23 in chapter 1.

Page 83—**value of a statistical life**: For two good recent surveys, see Ashenfelter, "Measuring the Value of a Statistical Life," and Viscusi and Aldy, "Value of a Statistical Life."

Page 83—**one percent chance**: For the full quotation and the argument for why this is a false equivalency, see Sunstein, *Worst-Case Scenarios*. Whatever damage the certain event would cause, having it occur with a probability of 1 percent would imply that its damage estimates need to be divided by a hundred to have a sensible metric of comparison.

Page 84—**Worst-Case Scenarios**: For a discussion of worst-case scenarios, start with Sunstein, *Worst-Case Scenarios*. Another oft-cited treatise on the subject is Posner, *Catastrophe: Risk and Response*. See Parson, "The Big One," for a comprehensive summary and critique, as well as further important elaborations. For an attempt at a comprehensive classification that goes beyond our list of eight potential existential risks, see Bostrom and Ćirković, *Global Catastrophic Risks*. For a more technical discussion, with a "can-do" attitude, see Garrick, *Quantifying and Controlling Catastrophic Risks*.

Page 85—**Their verdict**: For a summary, see Parson, "The Big One."

Page 86—**underestimating the likelihood**: See "1-in-1,000-year event" on page 1 in chapter 1.

Page 86—**$2 to 3 billion**: See "$2 or 3 billion" on page 1 in chapter 1.

Page 87—**guiding principle**: See, among many others, Revesz and Livermore, *Retaking Rationality*.

Page 88—**nuclear terrorism is worse**: Bostrom and Ćirković, *Global Catastrophic Risks*, puts the probability of catastrophic nuclear terrorism at 1 to 5 percent. By contrast—and on the extreme end of educated guesses—Allison, *Nuclear Terrorism*, p. 15, states that "a nuclear terrorist attack on America in the decade ahead is more likely than not." Silver, "Crunching the Risk Numbers," converts this to a 5 percent chance per year of such a catastrophe striking this coming decade.

Page 90—**Contrast the historical precedent:** This logic and some language in this paragraph is taken from Weitzman, "Modeling and Interpreting the Economics."

CHAPTER 5. BAILING OUT THE PLANET

Page 92—**Bailing Out the Planet:** The technical term for the type of geoengineering we discuss here is "solar-radiation" or "shortwave radiation management," both abbreviated by "SRM." That stands in contrast to "direct carbon removal" (DCR) or "carbon dioxide removal" (CDR). (For the latter, see "comes under various guises" on page 107 in chapter 5 as well as the "Bathtub" entry on page 30 in chapter 2.)

Despite the science still being in its infancy, geoengineering is slowly but surely entering public conversations. For one of the best such public documents, see Keith, *A Case for Climate Engineering.* For one of the most accessible, see Goodell, *How to Cool the Planet.* For one of the strongest, well-argued rejoinders, see Hamilton, *Earthmasters.* For our own earlier take, see Wagner and Weitzman, "Playing God."

Page 92—**make development sustainable:** United Nations, *Our Common Future,* commonly known as the "Brundtland Report."

Page 92—**the Earth's atmosphere:** Global average temperatures increased 0.45°C from the average measured between 1861 and 1880 to the average measured between 1980 and 1989 (chapter 7, "Observed Climate Variations and Change," of the *IPCC First Assessment Report*).

Page 93—**displaced over 200,000:** McCormick, Thomason, and Trepte, "Atmospheric Effects."

Page 93—**tons of sulfur dioxide:** Estimates range from 17 million tons of sulfur dioxide (Self et al., "Atmospheric Impact") to above 20 million (Bluth et al., "Global Tracking"). Note that the metric here is sulfur dioxide. For the weight of sulfur alone, divide these estimates by two.

Page 93—**tons of carbon dioxide:** We calculated these numbers using parts per million (ppm) levels from Keeling et al., *Exchanges of Atmospheric CO_2,* and used the conversion rate of 2.13 billion tons of carbon per ppm from the Carbon Dioxide Information Analysis Center. ("Conversion Tables"). The generally accepted preindustrial level of carbon dioxide in the atmosphere is 280 ppm, or 2.19 trillion tons of carbon dioxide. In 1990, the measured carbon dioxide concentration was 355 ppm, equivalent to 2.77 trillion tons. That is 585 billion tons above the preindustrial level. By now, average carbon dioxide levels are 400 ppm, or around 3.1 trillion tons carbon dioxide. Subtracting, we get 940 billion tons above preindustrial levels.

Page 93—**still pointing up:** See "2 ppm" on page 22 in chapter 1.

Page 94—**5,000 times:** Little Boy was about 20,000 as powerful as traditional bombs at the time (White House Press Release on August 6, 1945). The power-to-mass ratio of Little Boy compared to one ton of conventional explosives averages out to about 4,500. The bomb killed over 80,000 people, even though only 1.38 percent of the bomb's nuclear core fissioned during the explosion (Schlosser, *Command and Control*). The most powerful atomic bomb deployed was Ivy King, with a power-to-mass ratio of roughly 128,000—a TNT-equivalent power of 500,000 tons, and weighing 3.9 tons ("Operation Ivy").

Page 94—**Titan II missile:** Eric Schlosser's *Command and Control* provides a terrific journalistic account of the evolution of the nuclear bomb and the ways the world has tried—and in some cases, almost failed—to control it.

Page 94—**a million to one:** Keith, *A Case for Climate Engineering*, p. 67, compares total tons of carbon dioxide to the effect of pumping one million tons of sulfur into the stratosphere every year. The resulting leverage ratio is near a million to one.

Page 95—**2 to 3 percent:** Mount Pinatubo was the best-studied volcanic eruption, with dozens of papers estimating total solar radiation impacts alone. Most present the results in Watts per square meter. Direct solar radiation decreased by as much as 25–30 percent as a direct result of the volcanic eruption. Averaged over the first 10 months, "monthly-mean clear-sky total solar irradiance at Mauna Loa, Hawaii, decreased by as much as 5 percent and averaged . . . 2.7 percent" (Dutton and Christy, "Solar Radiative Forcing"). Models later found similar results (Stenchikov et al., "Radiative Forcing"). The NASA Earth Observatory confirms these numbers: "While overall solar radiation was reduced by less than five percent, data showed a reduction of direct radiation by as much as 30 percent."

Page 95—**more acidic:** Caldeira and Wickett, "Oceanography." See also "Ocean acidification" on page 42 in chapter 2.

Page 95—**created more:** Incidentally, making oceans (or other ecosystems) even more acidic does not appear to be one of these problems. Carbon dioxide turns oceans more acidic. So does sulfur—in form of sulfuric acid—after it washes out of the atmosphere. However, acidification from carbon dioxide in oceans is 100 times as strong as any effects of sulfur deposition from Mount Pinatubo–style geoengineering, at least via the damage pathway of acid rain. Kravitz et al., "Sulfuric Acid Deposition," argues that "the additional sulfate deposition that would result from geoengineering will not be sufficient to negatively impact most ecosystems, even under the assumption that all deposited sulfate will be in the form of sulfuric acid."

Page 95—**low levels of stratospheric ozone**: McCormick, Thomason, and Trepte, "Atmospheric Effects," estimate that the eruption of Mount Pinatubo could have been responsible for a decrease of columnar ozone above the equator by 6–8 percent. Self et al., "Atmospheric Impact," show how the depletion of ozone after the eruption was higher than ever before recorded. Heckendorn et al., "Impact of Geoengineering Aerosols," use the ozone depletion associated with the Pinatubo eruption as a case study for their conclusion that geoengineering with tiny sulfur-based particles would result in a "significant depletion of the ozone layer."

The direct effects of Mount Pinatubo, however, should not be conflated with the overall effects of any future geoengineering efforts, as global temperature increases themselves may accelerate ozone destruction, an effect possibly reversed or prevented by geoengineering. See, for example, Kirk-Davidoff et al., "Effect of Climate Change," and Keith, "Photophoretic Levitation."

Page 95—**global dry spell**: Trenberth and Dai, "Effects of Mount Pinatubo." See also Jones, Sparks, and Valdes, "Supervolcanic Ash Blankets."

Page 96—**where should we stop**: Alan Robock cites this question of who controls the thermostat, among 19 other practical problems, as a reason geoengineering might be more trouble than it's worth. Robock, "20 Reasons." Another set of questions with no easy answers concerns the morality of "hacking the planet." Stephen Gardiner outlines a moral argument against geoengineering, and particularly against the idea that researching geoengineering is a "lesser evil" when compared to catastrophic climate change, in "Arming the Future."

Page 97—**subsidized worldwide**: See "$500 billion per year" on page 22 in chapter 1.

Page 98—**roughly a ton**: There are plenty of other direct and indirect effects of aviation. For comprehensive surveys, see Dorbian, Wolfe, and Waitz, "Climate and Air Quality Benefits," and Barrett, Britter, and Waitz, "Global Mortality."

Page 98—**$40 worth of damages**: See "possibly much more" on page 23 in chapter 1.

Page 98—**transatlantic flights**: A roundtrip flight from New York City to Europe has a carbon footprint of 2–3 tons per passenger. Rosenthal, "Biggest Carbon Sin."

Page 98—**30 million; three billion**: Numbers for 2012 from International Civil Aviation Organization (ICAO)'s "The World of Civil Aviation: Facts and Figures."

Page 99—**Voluntary coordination**: The late Ronald Coase, the originator of the idea that—under certain strong conditions—coordination among individuals ("Coasian bargaining") can arrive at the socially optimal solution, would have agreed. See Glaeser, Johnson, and Shleifer, "Coase vs. the Coasians." One major stumbling block is the presence of large transaction costs for a negotiation among so many actors. Coase is widely credited with introducing that very idea of transaction costs to economics, which he used to explain the role of firms (Coase, "The Nature of the Firm"). The seminal article that introduced what would later be known as "Coasian bargaining" made it clear that well-defined property rights and low transaction costs were a precondition for its success (Coase, "The Problem of Social Cost").

Page 100—**too cheap to ignore**: Royal Society, "Geoengineering the Climate," estimates the cost of cooling the planet through tiny particles injected into the stratosphere to be $0.2 billion/year/W/m^2. This is compared to an estimated $200 billion/year/W/m^2 for reducing carbon dioxide in the first place. Schelling, "Economic Diplomacy of Geoengineering," is among the first economists to make this point. Barrett, "Incredible Economics of Geoengineering," may be the most prominent. Keith, "Geoengineering the Climate," and Royal Society, "Geoengineering the Climate," are the most authoritative. Goes, Tuana, and Keller, "Economics (or Lack Thereof)," and Klepper and Rickels, "Real Economics of Climate Engineering," have since added important caveats. McClellan, Keith, and Apt, "Cost Analysis," has recently added further perspectives. Finally, Bickel and Agrawal, "Reexamining the Economics," extends the work of Goes et al. and changes some assumptions to find that geoengineering will pass a benefit-cost test under more scenarios.

Page 102—**Asilomar Process**: Berg, "Asilomar and Recombinant DNA." For the original Asilomar statement, see Berg et al., "Summary Statement."

Page 102—**last of these headlines**: Giles, "Hacking the Planet."

Page 102—**Asilomar 2.0.**: Environmental Defense Fund was one of the cosponsors of the event.

Page 103—**firsthand account:** Schneider, *Science as a Contact Sport.*

Page 103—**The final statement:** "Asilomar Conference Recommendations," prepared by the Asilomar Scientific Organizing Committee.

Page 103—**Mount Pinatubo–style remedies:** Geoengineering has garnered a significant amount of attention since the 2006 publication of Crutzen, "Albedo Enhancement," which broke a long-standing taboo of sorts. An informal survey of the 77 articles on "geoengineering" in the journal *Climatic Change* shows that 19 had been published in the 18 years from 1977 to 2005. Between 2006 and 2013, the number was 58. The year 2013 alone saw the publication of 23 articles on geoengineering, and that's just in this one journal.

Page 104—**trade-off:** Many economists call this "moral hazard," which David Keith may have been the first to use in the geoengineering context (Keith, "Geoengineering the Climate"). The label has stuck, even though Scott Barrett has argued convincingly that it isn't technically true. Moral hazard refers to incentive problems between two parties. Driving faster because of wearing a seat belt is simply lack of self-control. Similarly, Keith, *A Case for Climate Engineering*, p. 139, describes some of the ensuing debate as "moral confusion, not moral hazard."

Page 105—**vast majority of Americans:** Tony Leiserowitz presented these results at the Asilomar Conference in March 2010. He has not asked the question since.

Page 105—**Painting roofs white:** See Menon et al., "Radiative Forcing." The *Royal Society*'s "Geoengineering the Climate" describes roof whitening as one of the "least effective and most expensive methods considered." The report estimates roof whitening to be 10,000 times more expensive per W/m^2 reduction in radiative forcing than Mount Pinatubo–style geoengineering.

Page 105—**vicious circles:** Curry, Schramm, and Ebert, "Sea Ice-Albedo."

Page 106—**urban areas elsewhere:** Oleson, Bonan, Feddema. "Effects of White Roofs," finds that painting roofs white in urban setting could reduce the urban heat island effect by a third, reducing daily maximum temperature by 0.6°C (1.1°F).

Page 106—**making things worse:** Jacobson and Ten Hoeve, "Urban Surfaces and White Roofs."

Page 106—**a tenth of the impact:** Menon et al., "Radiative Forcing," estimates the carbon dioxide offset of painting all roofs and pavements in urban areas white to be around 57 billion tons. Mount Pinatubo's eruption offset 585 billion tons of carbon dioxide.

Page 106—**need for air-conditioning:** "Cool Roof Fact Sheet."

Page 106—**some serious proposals**: Several recent studies look at the effects. See, for example, Latham et al., "Marine Cloud Brightening," Jones, Haywood, and Boucher, "Geoengineering Marine Stratocumulus Clouds," Latham et al., "Global Temperature Stabilization," Salter, Sortino, and Latham, "Sea-Going Hardware."

Page 107—**Indian monsoon**: Keith, *A Case for Climate Engineering*, p. 57–60, describes discussion of the Indian monsoon as among the most polarizing regional effects of geoengineering in the context of injecting sulfates into the stratosphere. Compare, for example, Robock, Oman, and Stenchikov, "Regional Climate Responses," with Pongratz et al., "Crop Yields." The former points to geoengineering as potentially "reducing precipitation to the food supply for billions of people." The latter points to geoengineering as potentially increasing crop yields in India.

Page 107—**comes under various guises**: See the Royal Society's "Geoengineering the Climate" for a comprehensive overview of geoengineering methods. All come with their own caveats and exceptions. The efficacy of some is under serious dispute. One recent study on biochar, for example, shows that it might not work as well as previously thought. Jaffé et al., "Global Charcoal Mobilization," finds that the carbon is not all captured, but rather a large portion dissolves and is released into rivers and oceans. Multiple other studies show a range of estimates for the "mean residence time" of biochar, ranging from 8.3 years (Nguyen et al., "Long-term Black Carbon"), to 3,624 years (Major et al., "Fate of Soil-Applied Black *Carbon*). Gurwick et al., "Systematic Review of Biochar Research," reviewed over 300 peer-reviewed articles on biochar and concluded that it's impossible to conclude very much at all based on the limited and wide range of data currently available.

Page 108—**Ocean fertilization**: Many scientists think that ocean fertilization is an inefficient route to carbon removal, and implementation on a large scale would likely be ineffective and disruptive to the marine ecosystem. Strong et al., "Ocean Fertilization."

Page 110—**0.8°C**: See "warmed by 0.8°C (1.4°F)" on page 13 in chapter 1.

Page 110—**By 2100**: The Summary for Policymakers of the *IPCC Fifth Assessment Report*'s Working Group I gives 3–5°C as the approximate range of temperature change by 2100 for the RCP8.5 scenario. The U.S. EPA estimates temperature changes up to 11.5°F by 2100 ("Future Climate Change").

Page 110—**serious problems**: See "Mark Lynas" and "HELIX" on page 14 in chapter 1.

Page 110—**A sudden jump**: The technical term is the "termination effect." Jones et al., "Impact of Abrupt Suspension," used 11 different climate models to examine this effect. They found substantial agreement among the models that sudden termination of long-term geoengineering would induce rapid increase in mean global temperature and precipitation, as well as a rapid decrease in sea ice cover. Matthews and Caldeira, "Transient Climate–Carbon Simulations," estimated that warming rates after a sudden termination of geoengineering could be up to 20 times those today.

Page 110—**national security threat**: See Gwynne Dyer's *Climate Wars* for one of the most vivid takes. The "Quadrennial Defense Review Report" from the U.S. Department of Defense declared, "climate change and energy are two key issues that will play a significant role in shaping the future security environment." Hsiang, Meng, and Cane, "Civil Conflicts," shows just that in the historical record, demonstrating that El Niño / Southern Oscillation may have played a role in a fifth of all civil conflicts since 1950. Hsiang, Burke, and Miguel, "Influence of Climate" reviews 60 studies on climate and human conflict and finds a substantial causal link between the two.

Page 111—**means less rainfall**: For a good survey of this phenomenon, see Ricke, Morgan, and Allen, "Regional Climate Response." Self et al., "Atmospheric Impact," note that the Mississippi floods could be attributable to the Mount Pinatubo eruption. See also Christensen and Christensen, "Climate Modelling." For more on general attribution science around climate change rather than geoengineering, see "attribution science" on page 4 in chapter 1.

Page 111—**attribution science**: See "attribution science" on page 4 in chapter 1.

Page 112—**Commission is worse**: See Samuelson and Zeckhauser, "Status Quo Bias," and Kahneman, Knetsch, and Thaler, "Anomalies." For a closely related concept, the "doctrine of double effect," see Thomson, "The Trolley Problem." Also see "errors of omission become as bad" on page 125.

Page 112—**best-studied**: McCormick, Thomason, and Trepte, "Atmospheric Effects."

Page 112—**experiment with the atmosphere**: See Robock, "Is Geoengineering Research Ethical?," for an ethical argument against geoengineering research outside the lab. There is indeed a broader set of issues, sometimes referred to as the "Collingridge dilemma": we can't know about the impacts of a technology until we have it; and once we have it, basic forces push us toward using it (Collingridge, *The Social Control of Technology*).

Page 113—**since the 1800s**: See the "Climate Science" entry on page 35 in chapter 2.

Page 113—**term "global warming"**: See "Wally Broecker" on page 49 in chapter 3.

Page 114—**other similar efforts**: Asilomar 2.0 is only one example. Another is the Solar Radiation Management Governance Initiative, convened by the British Royal Society, the Academy of Sciences for the Developing World, and the Environmental Defense Fund. By some assessments, Asilomar itself was only an extension of the "Oxford Principles" on geoengineering. These principles were submitted in 2009 to the UK House of Commons Science and Technology Select Committee's report, "The Regulation of Geoengineering," and subsequently endorsed by both the committee and the UK government. The authors of the principles also wrote a paper explaining their function and proposing a method for their implementation. (See Rayner et al., "The Oxford Principles.")

Page 114—**End the Deadlock**: Parson and Keith, "End the Deadlock." This isn't David Keith's first foray into governance issues by far. See http://www.keith.seas.harvard.edu/geo-engineering/.

CHAPTER 6. 007

Page 120—**possibility of a "Greenfinger"**: Wood, "Re-engineering the Earth." In fact, some might say it has already happened, at least on a tiny scale. In 2012, adversaries of geoengineering and scientists alike were incensed when they discovered that American businessman Russ George had conducted a rogue "experiment" of ocean fertilization, dumping 100 tons of iron sulfate (five times more than any previous fertilization experiment) into the Pacific Ocean in order to spark an enormous plankton growth, which he thought would both suck carbon out of the atmosphere, and aid in the recovery of the local salmon fishery. George's "experiment" was attacked as unscientific, illegitimate, and irresponsible, and George himself was dubbed "the first geo-vigilante" (Specter, "The First Geo-Vigilante"; Fountain, "Rogue Climate Experiment"). It turns out that the fishing village Old Massett of Haida Nation had voted to lend money to the Haida Salmon Restoration Corporation for the project in the hopes of bringing the local salmon fishery back from the brink, and George was brought on as chief scientist only later. It is as of yet unclear if the experiment will help restore the salmon population (Tollefson, "Ocean-Fertilization").

Page 120—**other question marks:** For another take on a future scenario of a geoengineered planet, see Weitzman, "The Geoengineered Planet." For perhaps the most comprehensive take on the science, politics, and ethics—with a strong point of view—see (once again) Keith, *A Case for Climate Engineering.*

Page 120—**Tens of millions more:** From the major East Asian rivers, Brahmaputra and Indus are likely to be most affected by melting Himalayan glaciers, "threatening the food security of an estimated 60 million people." Immerzeel, van Beek, and Bierkens, "Asian Water Towers."

Page 123—**bad health effects:** The potential health impacts of sulfur deposition from geoengineering is an area that has not yet been studied extensively. Initial results from a study conducted by David Keith at Harvard and Sebastian Eastham at MIT indicate that stratospheric injection of tiny particles could cause up to several thousand deaths per year. Another issue, quite separate from the direct health effects of sulfur, is potential sulfur deposition in oceans and other ecosystems. See "created more" on page 95 in chapter 5 on this point.

Page 124—**killing over 3.5 million:** In "Ambient (Outdoor) Air Quality and Health," the World Health Organization estimates that outdoor air pollution from human activities (e.g., transport and power generation) kills 3.7 million people annually. Indoor air pollution kills another 3.3 million for a total of seven million people ("7 Million Premature Deaths Annually Linked to Air Pollution").

Page 124—**Avoiding blame:** Weaver, *Politics of Blame Avoidance.*

Page 125—**errors of omission become as bad:** This thought experiment has a well-grounded foundation in moral philosophy, with no good solution to speak of. It is a matter of degrees. See Parfit, "Five Mistakes in Moral Mathematics." The same question is often presented in the so-called trolley problem. See Michael Sandel's *Justice*, David Edmonds's *Would You Kill the Fat Man?*, and Thomson, "The Trolley Problem."

In *Reasons and Persons* Parfit also identifies another, oft-stated philosophical objection to worrying about the effects of climate change (as well as geoengineering) in the first place: the "nonidentity problem." Climate change will alter the course of history as we know it, changing human settlement, migration, and, thus, mating patterns. As a result, future generations will be made up entirely of people who would not have been born without the effects of climate change. How then can we say that future generations will be harmed by climate change (or geoengineering), if the exact same people would not even be alive without climate change (or geoengineering)? Parfit himself, quite rightfully, identifies the "non-identity problem" as something that merits an immediate workaround, and there are several. Perhaps the best for our purposes is that the act itself (climate change or geoengineering) is potentially bad for the future person without making him or her strictly worse off in the non-identity sense of the word of not having been born. Either way, the distinction around errors of commission and omission stands. And in some ways, the issue of errors of commission and omission by various degrees is significantly more difficult to resolve than the fundamental (non-) objection of the "non-identity problem."

Page 126—**The mathematical derivation**: For the technical derivations, see Weitzman, "Voting Architecture." The paper derives the ideal voting rule in terms of Type I and Type II errors. The technical definition of a Type I is the incorrect rejection of a particular hypothesis. Assume that climate change is so bad it requires a geoengineering intervention. Proceed accordingly, only to find out later that geoengineering does more harm than good: an error of commission. Type II errors correspond to errors of omission in this thought experiment: Assume that climate change does not warrant a geoengineering intervention, only to find out later that it was indeed necessary, but now it is too late.

For a critical discussion of this voting architecture and two further analyses of geoengineering governance, see Barrett, "Solar Geoengineering's Brave New World."

CHAPTER 7. WHAT YOU CAN DO

Page 128—**1 in 60 million**: Gelman, Silver and Edlin, "What Is the Probability."

Page 128—**fraction of a penny**: Brennan, *Ethics of Voting*, p. 19, calculates the precise number in this hypothetical example to be 4.77×10 to the $-2,650$th power: approximately zero.

Page 129—**folk theory of voting ethics**: Brennan, *Ethics of Voting*.

Page 130—*But Will the Planet Notice?*: Wagner, *But Will the Planet Notice?* A version of the main arguments appeared as an op-ed in the *New York Times*: Wagner, "Going Green but Getting Nowhere."

Page 131—**if everyone does a little**: Emphasis not needed. David MacKay has italicized these words for us in: MacKay, *Sustainable Energy—without the Hot Air*.

Page 131—**self-perception theory**: Bem, "Self-Perception Theory." Also see Thøgersen and Crompton, "Simple and Painless?" for a comprehensive survey that points to the theories for complementarity from individual to collective action, and then points to the limitations of such a spillover.

Page 132—**bike to work**: See, for example: "Bike City," "Copenhagen: Bike City for More Than a Century," and "Bicycling History," *Cycling Embassy of Denmark.*

Page 132—**environmental decade**: See "Nixon went on to sign" on page 20 in chapter 1, as well as the text around it.

Page 133—**the "crowding-out bias"**: Another version of it is "single-action bias." Columbia's Center for Research on Environmental Decisions' CRED Guide provides an excellent resource on the psychology of climate change (communication) in general, in addition to a good primer on the single-action bias.

Page 133—**poorly studied**: Some research is beginning to investigate the link from individual to collective action and shows a self-reinforcing link, but only in a stated-preference context (Willis and Schor, "Changing a Light Bulb"). This type of research makes economists inherently uncomfortable. Asking people how they will act is one thing. Observing them is quite another.

Page 133—**crowds out virtuous behavior**: Titmuss's *The Gift Relationship* was among the first to hypothesize about this "crowding out" phenomenon from collective to individual action. Frey and Oberholzer-Gee, "Cost of Price Incentives," have revived interest in this work by establishing the theoretical underpinnings. Others have demonstrated its partial empirical validity, most notably perhaps in the context of paying for blood donations (Mellström and Johannesson, "Crowding Out in Blood Donation").

Page 133—**increase their electricity consumption**: The overall effect in terms of decreased emissions is still positive in this one example, as increased electricity use does not entirely offset the decreased pollution from participating in the program in the first place. See Jacobsen, Kotchen, and Vandenbergh, "Behavioral Response."

Page 136—**Sir Richard Branson**: Sir Richard Branson, chairman of Virgin Airlines, speaking at a U.S. State Department Conference on the "Global Impact Economy," on April 26, 2012 ("Interview of Virgin Group Ltd Chairman Sir Richard Branson by The Economist New York Bureau Chief Matthew Bishop").

Page 139—**two "100-year" storms**: See "Irene killed 49" and "Sandy killed 147" on page 2 in chapter 1.

Page 139—**price on carbon:** See "subsidized worldwide" on page 97 in chapter 5.

Page 140—**insurers and re-insurers:** Leurig and Dlugolecki, *Insurer Climate Risk*, offers a word of caution: Smaller insurers, in particular, may well need to be better prepared to weather their own climate risks.

Page 140—**rebuilding properties:** WNYC and ProPublica analyzed federal data and found that over 10,000 homes and business owners will be receiving Small Business Administration disaster loans to rebuild in flood-prone areas (Lewis and Shaw, "After Sandy"). New York has allocated $171 million to the buyout program, out of a $51 billion federal aid package. However, many homeowners are opting to rebuild in flood-prone zones rather than move to a new area (Kaplan, "Homeowners").

Page 141—**breaching typical New York seawalls:** See "three to twenty years" on page 5 in chapter 1.

Page 141—**hundreds of billions of dollars:** The New York Department of Finance's Fiscal Year 2014 Tentative Assessment Roll estimates the value of properties to be $873.7 billion.

Page 142—**stated emissions reductions targets:** See "700 ppm" on page 14 in chapter 1.

Page 142—**global warming exceeding 6°C:** See table 3.1 in chapter 3.

Page 143—**carbon dioxide alone:** See "400 parts per million" on page 10 and "2 ppm" on page 22 in chapter 1.

Page 143—**atmospheric tub:** See "The Bathtub Problem" beginning on page 15 of chapter 1 and the "Bathtub" entry on page 30 in chapter 2.

Page 144—**Bill McKibben:** McKibben, "Global Warming's Terrifying New Math." For additional analysis, see Generation Foundation, "Stranded Carbon Assets." For a good summary, see "A Green Light."

Page 145—**outperforming the market:** Margolis, Elfenbein, and Walsh, "Does It Pay to Be Good," find a small, positive effect. Eccles, Ioannou, and Serafeim, "Corporate Culture of Sustainability," match up "high" with "low" sustainability companies to find a sizeable, positive effect. Conversely, fossil fuel companies seem to have underperformed of late relative to broad market averages (Litterman, "The Other Reason for Divestment"). Investment under uncertainty is an important topic in and of itself. Option value theory applied to decreasing emissions and to coping with and profiting from climate change is clearly an important avenue for further research.

Page 145—**tobacco stocks:** The Australian High Court decision in favor of upholding the Tobacco Plain Packaging Act (2011) in *British American Tobacco Australasia Limited and Ors v. The Commonwealth of Australia*. See "Tobacco Shares Fall on Australian Packaging Rule."

EPILOGUE: A DIFFERENT KIND OF OPTIMISM

Page 148—**What We Know**: See whatweknow.aaas.org. The direct quote comes from the American Association for the Advancement of Science (AAAS) Climate Science Panel's background document "What We Know." Also see Melillo, Richmond, and Yohe, "Climate Change Impacts in the United States," and Risky Business Project, *Risky Business*. The latter shows how dealing with climate change is largely a risk management problem.

Page 149—**$40 per ton**: See "possibly much more" on page 23 in chapter 1.

Page 149—*negative* **$15**: See "$500 billion per year" on page 22 in chapter 1.

Page 149—**0.3 to 1 meters**: See "0.3 to 1 meters" on page 5 in chapter 1.

Page 149—**20 meters**: See "Global average temperatures" on page 10 in chapter 1.

Page 149—**1-in-10 chance**: See Table 3.1 in chapter 3.

Page 150—**cutting the flow**: See "The Bathtub Problem" on page 15 of chapter 1 and the "Bathtub" entry on page 30 in chapter 2.

Page 150—**baked in**: See "decades of warming" and "centuries of sea-level rise" on page 9 in chapter 1.

Page 151—**independent goal**: See Piketty, *Capital*, for perhaps the most comprehensive, contemporary argument.

Page 151—**taxing the rich and filthy**: See Klein, "Capitalism vs. the Climate," for the original quote, as well as Wagner, "Naomi Klein," for a response. Klein, *This Changes Everything*, emphasizes her earlier arguments. The book's subtitle: "Capitalism vs. the Climate."

Bibliography

||||||||||||||||||||||||

"7 Million Premature Deaths Annually Linked to Air Pollution." World Health Organization (March 25, 2014). http://www.who.int/media centre/news/releases/2014/air-pollution/en/.

"10-Year Treasury Inflation-Indexed Security, Constant Maturity." FRED Economic Data (March 17, 2014). http://research.stlouisfed.org/fred2 /series/DFII10/.

Acemoglu, Daron, Philippe Aghion, Leonardo Bursztyn, and David Hemous. "The Environment and Directed Technical Change." *American Economic Review* 102.1 (2012): 131–66. http://dspace.mit.edu/open access-disseminate/1721.1/61749.

"Acid Test." *Economist* (November 23, 2013). http://www.economist.com /news/science-and-technology/21590349-worlds-seas-are-becoming -more-acidic-how-much-matters-not-yet-clear.

Alderman, Liz. "A Greener Champagne Bottle." *New York Times* (September 1, 2010): B1. http://www.nytimes.com/2010/09/01/business /energy-environment/01champagne.html.

Aldy, Joseph E., and William A. Pizer. "Comparability of Effort in International Climate Policy Architecture." The Harvard Project on Climate Agreements Discussion Paper 14–62 (January 2014). http:// belfercenter.ksg.harvard.edu/files/dp62_aldy-pizer.pdf.

Allison, Graham T. *Nuclear Terrorism: The Ultimate Preventable Catastrophe*. Times Books, 2004. http://books.google.com/books?id=jDFY6FY 4aakC&dq=isbn:0805076514.

Altshuller, A. P. "Assessment of the Contribution of Chemical Species to the Eye Irritation Potential of Photochemical Smog." *Journal of the Air Pollution Control Association* 28.6 (1978): 594–98. http://www.tandf online.com/doi/abs/10.1080/00022470.1978.10470634#.UfkVe STD99.

"Ambient (Outdoor) Air Quality and Health." World Health Organization (March 2014). http://www.who.int/mediacentre/factsheets /fs313/en/.

Anthoff, David, Robert J. Nicholls, Richard SJ Tol, and Athanasios T. Vafeidis. "Global and Regional Exposure to Large Rises in Sea-Level: A Sensitivity Analysis." Tyndall Centre for Climate Change Research Working Paper 96 (2006). http://www.tyndall.ac.uk/sites/default /files/wp96_0.pdf.

Anttila-Hughes, Jesse Keith and Solomon M. Hsiang. "Destruction, Disinvestment, and Death: Economic and Human Losses Following Environmental Disaster" (February 18, 2013). http://dx.doi.org/10.2139 /ssrn.2220501.

Archer, David, Michael Eby, Victor Brovkin, Andy Ridgwell, Long Cao, Uwe Mikolajewicz, Ken Caldeira et al. "Atmospheric Lifetime of Fossil Fuel Carbon Dioxide." *Annual Review of Earth and Planetary Sciences* 37 (2009): 117–34. http://forecast.uchicago.edu/Projects/archer .2009.ann_rev_tail.pdf.

Arrhenius, Svante. "On the Influence of Carbonic Acid in the Air upon the Temperature of the Ground." *London, Edinburgh, and Dublin Philosophical Magazine and Journal of Science* 41.251 (1896): 237–76. http://www.tandfonline.com/doi/abs/10.1080/14786449608620846# .Ufl0myTD99A.

———. *Worlds in the Making: The Evolution of the Universe.* Harper, 1908. http://books.google.com/books/about/Worlds_in_the_making.html.

Arrow, K., M. Cropper, C. Gollier, B. Groom, G. Heal, R. Newell, W. Nordhaus et al. "Determining Benefits and Costs for Future Generations." *Science* 341.6144 (2013): 349–50. http://www.sciencemag .org/content/341/6144/349.full.

Artemieva, Natalia. "Solar System: Russian Skyfall." *Nature* 503.7475 (2013): 202–3. http://www.nature.com/nature/journal/v503/n7475 /full/503202a.html.

Ashenfelter, Orley. "Measuring The Value of A Statistical Life: Problems and Prospects." *Economic Journal*, 116.510 (2006): C10–C23. http:// www.nber.org/papers/w11916.

"The Asilomar Conference Recommendations on Principles for Research into Climate Engineering Techniques: Conference Report," prepared by the Asilomar Scientific Organizing Committee, Climate Institute (2010). http://climateresponsefund.org/images/Conference /finalfinalreport.pdf.

Avila, Lixion A., and John Cangialosi. "Tropical Cyclone Report: Hurricane Irene." National Hurricane Center (December 2011). http:// www.nhc.noaa.gov/data/tcr/AL092011_Irene.pdf.

Axelrad, Daniel A., David C. Bellinger, Louise M. Ryan, and Tracey J. Woodruff. "Dose–Response Relationship of Prenatal Mercury Exposure and IQ: An Integrative Analysis of Epidemiologic Data." *Environmental Health Perspectives* 115. 4 (2007): 609. http://www.ncbi.nlm .nih.gov/pmc/articles/PMC1852694/.

Ballantyne, A. P., C. B. Alden, J. B. Miller, P. P. Tans, and J.W.C. White. "Increase in Observed Net Carbon Dioxide Uptake by Land and Oceans

During the Past 50 Years." *Nature* 488.7409 (2012): 70–72. http://www
.nature.com/nature/journal/v488/n7409/full/nature11299.html.

Bamber, Jonathan L., Riccardo E. M. Riva, Bert L. A. Vermeersen, and
Anne M. LeBrocq. "Reassessment of the Potential Sea-Level Rise from
a Collapse of the West Antarctic Ice Sheet." *Science* 324.5929 (2009):
901–3. http://www.sciencemag.org/content/324/5929/901.short.

Barreca, Alan, Karen Clay, Olivier Deschenes, Michael Greenstone,
and Joseph S. Shapiro. "Adapting to Climate Change: The Remarkable Decline in the U.S. Temperature-Mortality Relationship over
the 20th Century." NBER Working Paper No. 18692 (January 2013).
http://www.nber.org/papers/w18692.

Barrett, Scott. "Climate Treaties and Approaching Catastrophes." *Journal of
Environmental Economics and Management* 66.2 (2013): 235–50. http://
www.sciencedirect.com/science/article/pii/S0095069612001222.

———. *Environment and Statecraft: The Strategy of Environmental Treaty-
Making.* Oxford University Press, 2003. http://global.oup.com/academic
/product/environment-and-statecraft-9780199257331?cc=us&lang
=en&tab=reviews.

———. "The Incredible Economics of Geoengineering." *Environmental
and Resource Economics* 39 (2008): 45–54. http://link.springer.com/article
/10.1007%2Fs10640-007-9174-8.

———. "Solar Geoengineering's Brave New World: Thoughts on the
Governance of an Unprecedented Technology." *Review of Environmental Economics and Policy* 8.2 (2014): 249–69. http://reep.oxfordjournals
.org/content/8/2/249.abstract.

Barrett, Scott, and Astrid Dannenberg. "Climate Negotiations under
Scientific Uncertainty." *Proceedings of the National Academy of Sciences*
109.43 (2012): 17372–76. http://www.pnas.org/content/109/43/17372
.short.

———. "Sensitivity of Collective Action to Uncertainty about Climate
Tipping Points." *Nature Climate Change* 4.1 (2014): 36–39. http://
www.nature.com/nclimate/journal/v4/n1/full/nclimate2059.html.

Barrett, Steven R. H., Rex E. Britter, and Ian A. Waitz. "Global Mortality
Attributable to Aircraft Cruise Emissions." *Environmental Science &
Technology* 44.19 (2010): 7736–42. http://pubs.acs.org/doi/abs/10.1021
/es101325r.

Barro, Robert J. "Rare Disasters, Asset Price, and Welfare Costs." *American Economic Review* 99.1 (2009): 243–64. http://www.nber.org/papers
/w13690.

Bazilian, Morgan, Ijeoma Onyeji, Michael Liebreich, Ian MacGill, Jennifer Chase, Jigar Shah, Dolf Gielen, Doug Arent, Doug Landfear,

and Shi Zhengrong. "Re-considering the Economics of Photovoltaic Power." *Renewable Energy* 53 (2013): 329–38. http://www.science direct.com/science/article/pii/S0960148112007641.

Bem, Daryl. "Self-Perception Theory." In *Advances in Experimental Social Psychology*, edited by Leonard Berkowitz, vol. 6. Academic Press, 1972. 1–62.

Benedick, Richard Elliot. *Ozone Diplomacy: New Directions in Safeguarding the Planet*. Harvard University Press, 1998.

Benenson Strategy Group and GS Strategy Group. "Recent Polling on Youth Voters" (July 2013). http://www.lcv.org/issues/polling/recent -polling-on-youth.pdf.

Berg, Paul. "Asilomar and Recombinant DNA." *Nobelprize.org* (July 17, 2013). http://www.nobelprize.org/nobel_prizes/chemistry/laureates /1980/berg-article.html.

Berg, Paul, D. Baltimore, S. Brenner, R. O. Robin, and M. F. Singer. "Summary Statement of the Asilomar Conference on Recombinant DNA Molecules." *Proceedings of the National Academy of Sciences of USA* 72.6 (1975): 1981–84. http://www.ncbi.nlm.nih.gov/pmc/articles /PMC432675/pdf/pnas00049-0007.pdf.

Bickel, J. Eric, and Shubham Agrawal. "Reexamining the Economics of Aerosol Geoengineering." *Climatic Change* (2011): 1–14. http://link .springer.com/article/10.1007/s10584-012-0619-x.

"Bicycling History." Cycling Embassy of Denmark. http://www.cycling -embassy.dk/facts-about-cycling-in-denmark/cycling-history/.

"Bike City." *Visitcopenhagen.com*. http://www.visitcopenhagen.com/copen hagen/bike-city.

Black, Fischer, and Robert B. Litterman. "Global Portfolio Optimization." *Financial Analysts Journal* 48.5 (1992): 28–43. http://www .jstor.org/discover/10.2307/4479577?uid=2&uid=4sid=21102739 263593.

Blake, Eric S., Todd B. Kimberlain, Robert J. Berg, John P. Cangialosi, and John L. Beven II. *Tropical Cyclone Report: Hurricane Sandy*. National Hurricane Center (February 2013). http://www.nhc.noaa.gov /data/tcr/AL182012_Sandy.pdf.

Bluth, Gregg J. S., Scott D. Doiron, Charles C. Schnetzler, Arlin J. Krueger, and Louis S. Walter. "Global Tracking of the SO2 Clouds from the June, 1991 Mount Pinatubo Eruptions." *Geophysical Research Letters* 19.2 (1992): 151–54. http://so2.gsfc.nasa.gov/pdfs/Bluth _Pinatubo1991_GRL91GL02792.pdf.

Borgerson, Scott G. "The Coming Arctic Boom." *Foreign Affairs* (July/ August 2013). http://www.foreignaffairs.com/articles/139456/scott-g -borgerson/the-coming-arctic-boom.

Bostrom, Nick, and Milan M. Ćirković, eds. *Global Catastrophic Risks*. Oxford University Press, 2011. http://www.global-catastrophic-risks .com/book.html.

Bradsher, Keith, and David Barboza. "Pollution from Chinese Coal Casts a Global Shadow." *New York Times* (June 11, 2006): A1. http://www .nytimes.com/2006/06/11/business/worldbusiness/11chinacoal.html.

Brauer, Michael, G. Hoek, H. A. Smit, J. C. de Jongste, J. Gerritsen, D. S. Postma, M. Kerkhof, and B. Brunekreef. "Air Pollution and Development of Asthma, Allergy and Infections in a Birth Cohort." *European Respiratory Journal* 29.5 (2007): 879–88. http://erj.ersjournals.com /content/29/5/879.full.

Brennan, Jason. *The Ethics of Voting*. Princeton University Press, 2011. http://press.princeton.edu/titles/9704.html.

British American Tobacco Australasia Limited and Ors v. the Commonwealth of Australia, no. S389/2411 (High Court of Australia 2012). http://www.hcourt.gov.au/cases/case-s389/2011.

Broecker, Wallace S. "Climatic Change: Are We on the Brink of a Pronounced Global Warming?" *Science* 189.4201 (1975). http://www .sciencemag.org/content/189/4201/460.abstract.

Brown, P. G., J. D. Assink, L. Astiz, R. Blaauw, M. B. Boslough, J. Borovicka, N. Brachet et al. "A 500-Kiloton Airburst over Chelyabinsk and an Enhanced Hazard from Small Impactors." *Nature* 503.7475 (2013): 238–41. http://www.nature.com/nature/journal/v503/n7475 /full/nature12741.html.

Budescu, David V., Han-Hui Por, Stephen B. Broomell, and Michael Smithson. "The Interpretation of IPCC Probabilistic Statements around the World." *Nature Climate Change* 4.6 (2014): 508–12. http://www.nature .com/nclimate/journal/vaop/ncurrent/full/nclimate2194.html.

Butler, James H., and Stephen A. Montzka. "The NOAA Annual Greenhouse Gas Index (AGGI)." *NOAA Earth System Research Laboratory* (2013). http://www.esrl.noaa.gov/gmd/ccgg/aggi.html.

Caldeira, Ken, and Michael E. Wickett. "Oceanography: Anthropogenic Carbon and Ocean pH." *Nature* 425.6956 (2003): 365. http://www .nature.com/nature/journal/v425/n6956/full/425365a.html.

Carlowicz, M. "World of Change: Global Temperatures." NASA Earth Observatory (2010). http://earthobservatory.nasa.gov/Features/World OfChange/decadaltemp.php.

Chakravarty, Shoibal, Ananth Chikkatur, Heleen de Coninck, Stephen Pacala, Robert Socolow, and Massimo Tavoni. "Sharing Global CO_2 Emission Reductions among One Billion High Emitters." *Proceedings of the National Academy of Sciences* 106.29 (2009): 11884–88. http:// cmi.princeton.edu/research/pdfs/one_billion_emitters.pdf.

Charney, Jule G., Akio Arakawa, D. James Baker, Bert Bolin, Robert E. Dickinson, Richard M. Goody, Cecil E. Leith, Henry M. Stommel, and Carl I. Wunsch. "Carbon Dioxide and Climate: A Scientific Assessment." National Academy of Sciences, 1979. http://www.atmos .ucla.edu/~brianpm/download/charney_report.pdf.

"China's 12GW Solar Market Outstripped All Expectations in 2013." *Bloomberg New Energy Finance* (January 23, 2014). http://about. bnef.com/files/2014/01/BNEF_PR_2014-01-23_China_Investment -final.pdf.

Christensen, Jens H., and Ole B. Christensen. "Climate Modelling: Severe Summertime Flooding in Europe." *Nature* 421.6925 (2003): 805–6. http://www.nature.com/nature/journal/v421/n6925/abs/421805a.html.

Christidis, Nikolaos, Peter A. Stott, Adam A. Scaife, Alberto Arribas, Gareth S. Jones, Dan Copsey, Jeff R. Knight, and Warren J. Tennant. "A New HadGEM3-A Based System for Attribution of Weather and Climate-Related Extreme Events." *Journal of Climate* 26.9 (2013). http://journals.ametsoc.org/doi/abs/10.1175/JCLI-D-12-00169.1.

Clark, Pilita. "What Climate Scientists Talk about Now." *Financial Times* (August 2, 2013). http://www.ft.com/intl/cms/s/2/4084c8ee-fa36-11e2 -98e0-00144feabdc0.html#axzz2cQegckSd.

"Clean Energy: Calculations and References." *U.S. Environmental Protection Agency* (last updated April 2013). http://www.epa.gov/clean energy/energy-resources/refs.html.

Clements, Benedict J., David Coady, Stefania Fabrizio, Sanjeev Gupta, Trevor Serge, Coleridge Alleyne, and Carlo A. Sdralevich, eds. *Energy Subsidy Reform: Lessons and Implications*. International Monetary Fund, 2013. http://www.amazon.com/Energy-Subsidy-Reform -Lessons-Implications/dp/1475558112/.

"Climate Science: Vast Costs of Arctic Change." *Nature* 499.7459 (2013): 401–3. http://www.naturecom/nature/journal/v499/n7459 /full/499401a.html.

CO_2Now. "Accelerating Rise of Atmospheric CO_2." http://co2now.org /Current-CO2/CO2-Trend/acceleration-of-atmospheric-co2.html.

Coase, Ronald H. "The Nature of the Firm." *Economica* 4.16 (1937): 386– 405. http://onlinelibrary.wiley.com/doi/10.1111/j.1468-0335.1937.tb 00002.x/full.

———. "The Problem of Social Cost." *The Journal of Law & Economics* 3 (1960): 1. http://heinonline.org/HOL/LandingPage?collection =journals&handle=hein.journals/jlecono3&div=2&id=&page=.

Collingridge, David. *The Social Control of Technology*. Pinter, 1980. http:// www.amazon.com/Social-Control-Technology-David-Collingridge /dp/031273168X.

"Conversion Tables." Carbon Dioxide Information Analysis Center (last updated September 2012). http://cdiac.ornl.gov/pns/convert.html.

"Cool Roof Fact Sheet." *US Department of Energy Building Technologies* (August 2009). http://www1.eere.energy.gov/buildings/pdfs/cool_roof_fact_sheet.pdf.

"Copenhagen: Bike City for More Than a Century." *Denmark.dk.* http://denmark.dk/en/green-living/bicycle-culture/copenhagen-bike-city-for-more-than-a-century/.

Coumou, Dim, and Alexander Robinson. "Historic and Future Increase in the Global Land Area Affected by Monthly Heat Extremes." *Environmental Research Letters* 8 (2013). http://iopscience.iop.org/1748-9326/8/3/034018/article.

Coumou, Dim, Alexander Robinson, and Stefan Rahmstorf. "Global Increase in Record-Breaking Monthly-Mean Temperatures." *Climatic Change* 118.3–4 (2013): 771–82. http://link.springer.com/article/10.1007%2Fs10584-012-0668-1.

Cropper, Maureen L., Mark C. Freeman, Ben Groom, and William A. Pizer. "Declining Discount Rates." *American Economic Review: Papers and Proceedings* 104.5 (2014): 538–43. http://dx.doi.org/10.1257/aer.104.5.538.

Crutzen, Paul J. "Albedo Enhancement by Stratospheric Sulfur Injections: A Contribution to Resolve a Policy Dilemma?" *Climatic Change* 77.3 (2006): 211–20. http://link.springer.com/article/10.1007%2Fs10584-006-9101-y.

Cui, Ying, Lee R. Kump, Andy J. Ridgwell, Adam J. Charles, Christopher K. Junium, Aaron F. Diefendorf, Katherine H. Freeman, Nathan M. Urban, and Ian C. Harding. "Slow Release of Fossil Carbon during the Palaeocene-Eocene Thermal Maximum." *Nature Geoscience* 4.7 (2011): 481–85. http://www.nature.com/ngeo/journal/v4/n7/full/ngeo1179.html.

Curry, Judith A., Julie L. Schramm, and Elizabeth E. Ebert. "Sea Ice–Albedo Climate Feedback Mechanism." *Journal of Climate* 8.2 (1995): 240–47. http://journals.ametsoc.org/doi/abs/10.1175/1520-0442(1995)008%3C0240:SIACFM%3E2.0.CO%3B2.

Dales, John H. *Pollution, Property, and Prices: An Essay in Policy-Making and Economics.* University of Toronto Press, 1968.

"Defending Planet Earth: Near-Earth Object Surveys and Hazard Mitigation Strategies." National Academies Press, 2010. http://www.nap.edu/download.php?record_id=12842.

"Demoralized Trading: Stocks Break on Record Volume—Banking Support Starts Rally." *Wall Street Journal* (October 25, 1929).

Deschênes, Olivier, and Enrico Moretti. "Extreme Weather Events, Mortality, and Migration." *Review of Economics and Statistics* 91.4 (2009): 659–81. http://www.mitpressjournals.org/doi/pdf/10.1162/rest.91.4.659.

Diffenbaugh, Noah S., and Christopher B. Field. "Changes in Ecologically Critical Terrestrial Climate Conditions." *Science* 341.6145 (2013): 486–92. http://www.sciencemag.org/content/341/6145/486.full.

Dorbian, Christopher S., Philip J. Wolfe, and Ian A. Waitz. "Estimating the Climate and Air Quality Benefits of Aviation Fuel and Emissions Reductions." *Atmospheric Environment* 45.16 (2011): 2750–59. http://www.sciencedirect.com/science/article/pii/S1352231011001592.

Dreifus, Claudia. "Chasing the Biggest Story on Earth." *New York Times* (February 11, 2014): D5. http://www.nytimes.com/2014/02/11/science/the-sixth-extinction-looks-at-human-impact-on-the-environment.html.

Dutton, Ellsworth G., and John R. Christy. "Solar Radiative Forcing at Selected Locations and Evidence for Global Lower Tropospheric Cooling Following the Eruptions of El Chichón and Pinatubo." *Geophysical Research Letters* 19.23 (1992): 2313–16. http://onlinelibrary.wiley.com/doi/10.1029/92GL02495/abstract.

Dyer, Gwynne. *Climate Wars: The Fight for Survival as the World Overheats*. Scribe Publications, 2008. http://www.amazon.com/Climate-Wars-Fight-Survival-Overheats/dp/1851688145.

Eccles, Robert, Ioannis Ioannou, George Serafeim. "The Impact of a Corporate Culture of Sustainability on Corporate Behavior and Performance." HBS working paper 12–035 (November 25, 2011). http://www.hbs.edu/research/pdf/12-035.pdf.

"Economics of Ocean Acidification." International Atomic Energy Agency workshop (November 2012). http://medsea-project.eu/wp-content/uploads/2013/10/ebook-Economics-of-Ocean-Acidification.pdf.

Edmonds, David. *Would You Kill the Fat Man? The Trolley Problem and What Your Answer Tells Us about Right and Wrong*. Princeton University Press, 2013. http://press.princeton.edu/titles/10074.html.

Ellerman, Denny, Frank Convery, and Christian de Perthuis. *Pricing Carbon: The European Union Emissions Trading Scheme*. Cambridge University Press, 2010. http://www.cambridge.org/us/academic/subjects/economics/natural-resource-and-environmental-economics/pricing-carbon-european-union-emissions-trading-scheme.

Emanuel, Kerry A. "Downscaling CMIP5 Climate Models Shows Increased Tropical Cyclone Activity over the 21st Century." *Proceedings of the National Academy of Sciences* (July 8, 2013). http://www.pnas.org/content/early/2013/07/05/1301293110.abstract.

——. "Increasing Destructiveness of Tropical Cyclones over the Past 30 Years." *Nature* 436.7051 (2005): 686–88. http://www.nature.com /nature/journal/v436/n7051/abs/nature03906.html.

——. "MIT Climate Scientist Responds on Disaster Costs and Climate Change." FiveThirtyEight (March 31, 2014). http://fivethirty eight.com/features/mit-climate-scientist-responds-on-disaster-costs -and-climate-change/.

Engber, Daniel. "You're Getting Warmer . . ." *Slate* (original story, 2007; reposted on August 20, 2013). http://www.slate.com/articles/health _and_science/science/2007/02/youre_getting_warmer_.html.

Fasullo, John T., and Kevin E. Trenberth. "A Less Cloudy Future: The Role of Subtropical Subsidence in Climate Sensitivity." *Science* 338.6108 (2012): 792–94. http://www.sciencemag.org/content /338/6108/792.

Fiscal Year 2014 Tentative Assessment Roll. NYC Finance (January 15, 2013). http://www.nyc.gov/html/dof/downloads/pdf/press_release/pr _assessment_14.pdf.

Fischer, Carolyn, Winston Harrington, and Ian WH Parry. "Should Automobile Fuel Economy Standards Be Tightened?" *Energy Journal* 28.4 (2007): 1–29. http://iaee.org/documents/vol28_4.pdf.

Fischetti, Mark. "New York City and the U.S. East Coast Must Take Drastic Action to Prevent Ocean Flooding." *Scientific American* 308.6 (2013): 58–67. http://www.scientificamerican.com/article.cfm?id=new-york -city-east-coast-drastic-actions-prevent-flooding-hurricane-sandy.

Fountain, Henry. "A Rogue Climate Experiment Outrages Scientists." *New York Times* (October 19, 2012). http://www.nytimes.com /2012/10/19/science/earth/iron-dumping-experiment-in-pacific- alarms-marine-experts.html.

Fourier, Jean-Baptiste Joseph. "Les Temperatures du Globe Terrestre et des espaces planetaires." *Mémoires de l'Académie des sciences de l'Institut de France* 7 (1827): 569–604. http://www.math.umn.edu/~mcgehee /Seminars/ClimateChange/references/Fourier1827.pdf.

——. "Remarques générales sur les températures du globe terrestre et des espaces planétaires." *Annales de Chimie et de Physique* 27 (1824): 136–67.

Fox-Penner, Peter. *Smart Power: Climate Change, the Smart Grid, and the Future of Electric Utilities.* Island Press, 2010. http://www.smartpower book.com/.

Franke, Andreas. "Analysis: German 2013 Wind, Solar Output Up 4% at Record 77TWh." *Platts* (January 7, 2014). http://www.platts.com/ latest-news/electric-power/london/analysis-german-2013-wind-solar -power-output-26598276.

Frey, Bruno S., and Felix Oberholzer-Gee. "The Cost of Price Incentives: An Empirical Analysis of Motivation Crowding Out." *American Economic Review* 87:746–55 (1997). http://www.jstor.org/stable/2951373.

Friedlingstein, P., R. M. Andrew, J. Rogelj, G. P. Peters, J. G. Canadell, R. Knutti, G. Luderer, M. R. Raupach, M. Schaeffer, D. P. van Vuuren, and C. Le Quéré. "Persistent Growth of CO_2 Emissions and Implications for Reaching Climate Targets." *Nature Geoscience* 7.10: 709–15 (2014). http://www.nature.com/ngeo/journal/vaop/ncurrent/full/ngeo2248.html.

"Future climate change." U.S. EPA. http://www.epa.gov/climatechange/science/future.html.

Gardiner, Stephen. "Is 'Arming the Future' with Geoengineering Really the Lesser Evil? Some Doubts about the Ethics of Intentionally Manipulating the Climate System." In *Climate Ethics: Essential Readings*, edited by Stephen M. Gardiner, Simon Caney, Dale Jamieson, and Henry Shue. Oxford University Press, 2010. 284–312. http://papers.ssrn.com/sol3/papers.cfm?abstract_id=1357162.

Garrick, B. John. *Quantifying and Controlling Catastrophic Risks*. Academic Press, 2008. http://www.amazon.com/Quantifying-Controlling-Catastrophic-Risks-Garrick/dp/0123746019.

Gelman, Andrew, Nate Silver, and Aaron Edlin. "What Is the Probability Your Vote Will Make a Difference?" *Economic Inquiry* 50.2 (2012): 321–26. http://www.stat.columbia.edu/~gelman/research/published/probdecisive2.pdf.

Generation Foundation. "Stranded Carbon Assets: Why and How Carbon Risks Should Be Incorporated in Investment Analysis." White Paper (October 30, 2013). http://genfound.org/media/pdf-generation-foundation-stranded-carbon-assets-v1.pdf.

Giles, Jim. "Hacking the Planet, Who Decides?" *New Scientist* (March 29, 2010). http://www.newscientist.com/article/dn18713-hacking-the-planet-who-decides.html?full=true#.UioxwyTD99A.

———. "Scientific Uncertainty: When Doubt Is a Sure Thing." *Nature* 418.6897 (2002): 476–78. http://www.nature.com/nature/journal/v418/n6897/full/418476a.html.

Gillett, Nathan P., Vivek K. Arora, Kirsten Zickfeld, Shawn J. Marshall, and William J. Merryfield. "Ongoing Climate Change Following a Complete Cessation of Carbon Dioxide Emissions." *Nature Geoscience* 4.2 (2011): 83–87. http://www.nature.com/ngeo/journal/v4/n2/abs/ngeo1047.html.

Gillingham, Kenneth, Richard G. Newell, and Karen Palmer. "Energy Efficiency Economics and Policy." *Annual Review of Resource Economics* 1

(2009): 597–620. http://www.annualreviews.org/doi/abs/10.1146/annu rev.resource.102308.124234.

Glaeser, Edward, Simon Johnson, and Andrei Shleifer. "Coase vs. the Coasians." *Quarterly Journal of Economics* 116.3 (2001): 853–99. http:// qje.oxfordjournals.org/content/116/3/853.short.

"Global Market Outlook for Photovoltaics 2013–2017." *European Photovoltaic Industry Association* (May 2013). http://www.epia.org/fileadmin /user_upload/Publications/GMO_2013_-_Final_PDF.pdf.

"Global Warming: Who Pressed the Pause Button." *Economist* (March 8, 2014). http://www.economist.com/news/science-and-technology/2159 8610-slowdown-rising-temperatures-over-past-15-years-goes-being.

Goes, Marlos, Nancy Tuana, and Klaus Keller. "The Economics (or Lack Thereof) of Aerosol Geoengineering." *Climatic Change* 109.3–4 (2011): 719–44. http://www.aoml.noaa.gov/phod/docs/Goes_etal_2011.pdf.

Gollier, Christian. "Evaluation of Long-Dated Investments under Uncertain Growth Trend, Volatility and Catastrophes." CESifo Working Paper Series No. 4052 (2012). http://papers.ssrn.com/sol3/papers.cfm ?abstract_id=2202094.

———. *Pricing the Planet's Future: The Economics of Discounting in an Uncertain World*. Princeton University Press, 2012. http://press.princeton .edu/titles/9894.html.

Gollier, Christian, and Martin L. Weitzman. "How Should the Distant Future Be Discounted When Discount Rates Are Uncertain?" *Economics Letters* 107.3 (2010): 350–53. http://scholar.harvard.edu/files /weitzman/files/howshoulddistantfuture.pdf.

Goodell, Jeff. *How to Cool the Planet: Geoengineering and the Audacious Quest to Fix Earth's Climate*. Houghton Mifflin Harcourt, 2010. http:// books.google.com/books?id=5hAnBB-wmH4C.

Goulder, Lawrence H., and Andrew R. Schein. "Carbon Taxes vs. Cap and Trade: A Critical Review." NBER Working Paper No. 19338 (August 2013). http://www.nber.org/papers/w19338.

Greenberg, Max, Denise Robbins, and Shauna Theel. "Media Sowed Doubt in Coverage of UN Climate Report." MediaMatters.org (October 10, 2013). http://mediamatters.org/research/2013/10/10/study -media-sowed-doubt-in-coverage-of-un-clima/196387.

"A Green Light." *Economist* (March 29, 2014). http://www.economist .com/news/business/21599770-companies-are-starting-open-up -about-their-environmental-risks-they-need-do-more-green.

Greenstone, Michael, Elizabeth Kopits, and Ann Wolverton. "Developing a Social Cost of Carbon for US Regulatory Analysis: A Methodology and Interpretation." *Review of Environmental Economics and Policy* 7.1 (2013): 23–46. http://reep.oxfordjournals.org/content/7/1/23.abstract.

Gunther, Marc. *Suck It Up: How Capturing Carbon from the Air Can Help Solve the Climate Crisis.* Amazon Kindle Single, 2012. http://www.marcgunther.com/suck-it-up-my-book-about-climate-change-geo engineering-and-air-capture-of-co2/.

Gurwick, Noel P., Lisa A. Moore, Charlene Kelly, and Patricial Elias. "A Systematic Review of Biochar Research, with a Focus on Its Stability in Situ and Its Promise as a Climate Mitigation Strategy." *Plos One* 8.9 (2013). http://www.plosone.org/article/info%3Adoi%2F10.1371%2Fjournal.pone.0075932.

Guy, Sophie, Yoshihisa Kashima, Iain Walker, and Saffron O'Neill. "Comparing the Atmosphere to a Bathtub: Effectiveness of Analogy for Reasoning about Accumulation." *Climatic Change* 121.4 (2013): 579–94. http://link.springer.com/article/10.1007/s10584-013-0949-3.

Haagen-Smit, A. J. "Chemistry and Physiology of Los Angeles smog." *Industrial and Engineering Chemistry* 44.6 (1952): 1342–46. http://pubs.acs.org/doi/abs/10.1021/ie50510a045?journalCode=iechad.

Hamilton, Clive. *Earthmasters: The Dawn of the Age of Climate Engineering.* Yale University Press, 2013. http://books.google.com/books?lr=&id=x61F2HkKtVEC.

Hammar, H., T. Sterner, and S. Åkerfeldt. "Sweden's CO_2 Tax and Taxation Reform Experiences." In *Reducing Inequalities: A Sustainable Development Challenge*, edited by R. Genevey, R. Pachauri, L., Tubiana. New Delhi: TERI Press (2013): 169–74. http://www.efdinitiative.org/publications/swedens-co2-tax-and-taxation-reform-experiences.

Hansen, James, and Makiko Sato. "Climate Sensitivity Estimated from Earth's Climate History." NASA Goddard Institute for Space Studies and Columbia University Earth Institute. New York (May 2012). http://www.columbia.edu/~jeh1/mailings/2012/20120508_Climate Sensitivity.pdf.

Hansen, James, Makiko Sato, Pushker Kharecha, David Beerling, Valerie Masson-Delmotte, Mark Pagani, Maureen Raymo, Dana L. Royer, and James C. Zachos. "Target Atmospheric CO_2: Where Should Humanity Aim?" *NASA Goddard Institute for Space Studies, Columbia University Earth Institute* (2008). http://www.columbia.edu/~jeh1/2008/TargetCO2_20080407.pdf.

Hardin, Garrett. "Tragedy of the Commons." *Science* 162.3859 (1968): 1243–48. http://citeseerx.ist.psu.edu/viewdoc/download?doi=10.1.1.124.3859&rep=rep1&type=pdf.

Harvey, L.D.D. "Mitigating the Atmospheric CO_2 Increase and Ocean Acidification by Adding Limestone Powder to Upwelling Regions." *Journal of Geophysical Research: Oceans (1978–2012)* 113.C4 (2008). http://onlinelibrary.wiley.com/doi/10.1029/2007JC004373/abstract.

Heal, Geoffrey M., and Antony Millner. "Agreeing to Disagree on Climate Policy." *Proceedings of the National Academy of Sciences* 111.10 (2014): 3695–8. http://www.pnas.org/content/early/2014/02/19/1315987111.short.

———. "Uncertainty and Decision in Climate Change Economics." NBER Working Paper No. 18929 (2013). http://www.nber.org/papers/w18929.

Heal, Geoffrey M., and Jisung Park. "Feeling the Heat: Temperature, Physiology and the Wealth of Nations." National Bureau of Economic Research (NBER) Working Paper No. w19725 (2013). http://www.nber.org/papers/w19725.

Heckendorn, P., D. Weisenstein, S. Fueglistaler, B. P. Luo, E. Rozanov, M. Schraner, L. W. Thomason, and T. Peter. "The Impact of Geoengineering Aerosols on Stratospheric Temperature and Ozone." *Environmental Research Letters* 4.4 (2009). http://iopscience.iop.org/1748-9326/4/4/045108/fulltext/.

Heffernan, Margaret. *Willful Blindness*. Walker, 2011. http://books.google.com/books?id=3rQiXitkUpMC.

High-End cLimate Impacts and eXtremes (HELIX). http://www.HELIXclimate.eu.

Hogan, William W. "Scarcity Pricing: More on Locational Operating Reserve Demand Curves" Harvard Electricity Policy Group. Presentation, San Diego, March 2013. http://www.hks.harvard.edu/fs/whogan/Hogan_hepg_031309r.pdf.

House of Commons Science and Technology Committee. "The Regulation of Geoengineering." London, UK (2010). http://www.publications.parliament.uk/pa/cm200910/cmselect/cmsctech/221/221.pdf.

Howard, Peter. "Omitted Damages: What's Missing from the Social Cost of Carbon." Cost of Carbon Project (2014). http://costofcarbon.org/files/Omitted_Damages_Whats_Missing_From_the_Social_Cost_of_Carbon.pdf.

Hsiang, Solomon M., Marshall Burke, and Edward Miguel. "Quantifying the Influence of Climate on Human Conflict." *Science* 341.6151 (2013): 1235367. http://www.sciencemag.org/content/341/6151/1235367.

Hsiang, Solomon M., Kyle C. Meng, and Mark A. Cane. "Civil Conflicts Are Associated with the Global Climate." *Nature* 476.7361 (2011): 438–41. http://www.nature.com/nature/journal/v476/n7361/full/nature10311.html.

IGBP, IOC, SCOR. "Ocean Acidification Summary for Policymakers—Third Symposium on the Ocean in a High-CO_2 World." International Geosphere-Biosphere Programme, Stockholm, Sweden (2013).

http://www.igbp.net/download/18.30566fc6142425d6c91140a
/1384420272253/OA_spm2-FULL-lorez.pdf.

Immerzeel, Walter W., Ludovicus P. H. van Beek, and Marc F. P. Bierkens. "Climate Change Will Affect the Asian Water Towers." *Science* 328.5984 (2010): 1382–85. http://www.sciencemag.org/content /328/5984/1382.full.

"Incorporating Sea-Level Change Considerations in Civil Works Programs." USACE EC 1165–2–212 (2011). http://www.flseagrant.org /wp-content/uploads/2012/02/USACE_SLR_policy_2011-2013.pdf.

Inman, Mason, "Carbon Is Forever." *Nature Reports Climate Change* 12 (2008). http://www.nature.com/climate/2008/0812/full/climate.2008 .122.html.

Intergovernmental Panel on Climate Change (IPCC). *Climate Change 1992: The Supplementary Report to the IPCC Scientific Assessment* (1992). http://www.ipcc.ch/publications_and_data/publications_ipcc _supplementary_report_1992_wg1.shtml.

——. *IPCC Fifth Assessment Report: Climate Change 2013* (AR5) (2013– 14). http://www.ipcc.ch/report/ar5/.

——. *IPCC First Assessment Report* (1990).

——. *IPCC Fourth Assessment Report: Climate Change 2007* (AR4)(2007). http://www.ipcc.ch/publications_and_data/ar4/syr/en/contents.html.

——. *IPCC Second Assessment Report: Climate Change 1995 (SAR)* (1995). http://www.ipcc.ch/pdf/climate-changes-1995/ipcc-2nd-assessment /2nd-assessment-en.pdf.

——. *IPCC Third Assessment Report: Climate Change 2001 (TAR)* (2001). http://www.grida.no/publications/other/ipcc_tar/.

——. *Special Report: Managing the Risks of Extreme Events and Disasters to Advance Climate Change Adaptation* (2012). http://ipcc-wg2.gov /SREX/images/uploads/SREX-All_FINAL.pdf.

——. *IPCC Special Report on Emissions Scenarios (SRES)* (2000). https:// www.ipcc.ch/pdf/special-reports/spm/sres-en.pdf.

"Interview of Virgin Group Ltd Chairman Sir Richard Branson by *The Economist* New York Bureau Chief Matthew Bishop," transcript from U.S. Department of State conference on the "Global Impact Economy," on April 20, 2012. http://www.state.gov/s/partnerships /releases/2012/189100.htm.

"Irene by the Numbers." *NOAA* (August 2011). http://www.noaa.gov /images/Hurricane%20Irene%20by%20the%20Numbers%20-%20 Factoids_V4_083111.pdf.

Jacobsen, Grant D., Matthew J. Kotchen, and Michael P. Vandenbergh. "The Behavioral Response to Voluntary Provision of an Environmental Public Good: Evidence from Residential Electricity Demand."

European Economic Review 56 (2012): 946–60. http://www.nber.org /papers/w16608.

Jacobsen, Mark R. "Evaluating U.S. Fuel Economy Standards in a Model with Producer and Household Heterogeneity." *American Economic Journal: Economic Policy* 5.2 (2013): 148–87. http://www.aeaweb.org /articles.php?doi=10.1257/pol.5.2.148.

Jacobson, Mark Z., and John E. Ten Hoeve. "Effects of Urban Surfaces and White Roofs on Global and Regional Climate." *Journal of Climate* 24.3 (2012): 1028–44. http://journals.ametsoc.org/doi/abs/10.1175 /JCLI-D-11-00032.1?journalCode=clim.

Jaffé, Rudolf, Yan Ding, Jutta Niggermann, Anssi V. Vähätalo, Aron Stubbins, Robert G. M. Spencer, John Campbell, and Thorsten Dittmar. "Global Charcoal Mobilization from Soils via Dissolution and Riverine Transport to the Oceans." *Science* 340.6130 (2013): 345–47. http://www.sciencemag.org/content/340/6130/345.full.

Jensen, Robert T., and Nolan H. Miller. "Giffen Behavior and Subsistence Consumption." *American Economic Review* 98.4 (2008): 1553. http://www.aeaweb.org/articles.php?doi=10.1257/aer.98.4.1553.

Jerrett, Michael, Richard T. Burnett, C. Arden Pope III, Kazuhiko Ito, George Thurston, Daniel Krewski, Yuanli Shi, Eugenia Calle, and Michael Thun. "Long-Term Ozone Exposure and Mortality." *New England Journal of Medicine* 360.11 (2009): 1085–95. http://www.nejm .org/doi/full/10.1056/NEJMoa0803894.

Johansson, Bengt. "Economic Instruments in Practice 1: Carbon Tax in Sweden." *Swedish Environmental Protection Agency* (2001). http://www .oecd.org/science/inno/2108273.pdf.

"John Tyndall." *NASA Earth Observatory.* http://earthobservatory.nasa .gov/Features/Tyndall/.

Jones, Andy, Jim Haywood, and Olivier Boucher. "Climate Impacts of Geoengineering Marine Stratocumulus Clouds." *Journal of Geophysical Research: Atmospheres (1984–2012)* 114.D10 (2009). http://online library.wiley.com/doi/10.1029/2008JD011450/abstract.

Jones, Andy, Jim M. Haywood, Kari Alterskjær, Olivier Boucher, Jason N. S. Cole, Charles L. Curry, Peter J. Irvine et al. "The Impact of Abrupt Suspension of Solar Radiation Management (Termination Effect) in Experiment G2 of the Geoengineering Model Intercomparison Project (GeoMIP)." *Journal of Geophysical Research: Atmospheres* 118.17 (2013): 9743–52. http://onlinelibrary.wiley.com/doi/10.1002/jgrd.50762/abstract.

Jones, Gregory V., Michael A. White, Owen R. Cooper, and Karl Storchmann. "Climate Change and Global Wine Quality." *Climatic Change* 73.3 (2005): 319–43. http://www.recursosdeenologia.com/docs/2005 /2005_climate_change_and_global_wine_quality.pdf.

Jones, Morgan T., Stephen J. Sparks, and Paul J. Valdes. "The Climatic Impact of Supervolcanic Ash Blankets." *Climate Dynamics* 29.6 (2007): 553–64. http://link.springer.com/article/10.1007%2Fs00382 -007-0248-7.

Joos, Fortunat and M. Bruno. "A Short Description of the Bern Model." (September 1996). http://www.climate.unibe.ch/~joos/model_descrip tion/model_description.html.

Joughin, Ian, Benjamin E. Smith, and Brooke Medley. "Marine Ice Sheet Collapse Potentially Under Way for the Thwaites Glacier Basin, West Antarctica." *Science* 344.6185 (2014): 735–38. http://www.sciencemag .org/content/344/6185/735.abstract.

Kahn, Matthew E. *Climatopolis: How Our Cities Will Thrive in the Hotter Future*. Basic Books, 2010. http://books.google.com/books/about /Climatopolis.html?id=nQjxjwEACAAJ.

Kahneman, Daniel. *Thinking, Fast and Slow*. New York: Farrar, Straus and Giroux, 2011. http://www.amazon.com/Thinking-Fast-Slow-Daniel -Kahneman/dp/0374533555.

Kahneman, Daniel, and Amos Tversky. "Prospect Theory: An Analysis of Decision under Risk." *Econometrica* 47.2 (1979): 263–92. http:// pages.uoregon.edu/harbaugh/Readings/GBE/Risk/Kahneman%20 1979%20E,%20Prospect%20Theory.pdf.

———. "Subjective Probability: A Judgment of Representativeness." *Cognitive Psychology* 3.3 (1972): 430–54. http://www.sciencedirect .com/science/article/pii/0010028572900163.

Kahneman, Daniel, Jack L. Knetsch, and Richard H. Thaler. "Anomalies: The Endowment Effect, Loss Aversion, and Status Quo Bias." *Journal of Economic Perspectives* 5.1 (1991): 193–206. http://econ.ucdenver .edu/beckman/Econ%204001/thaler-loss-aversion.pdf.

———. "Experimental Tests of the Endowment Effect and the Coase theorem." *Journal of Political Economy* 98.6 (1990): 1325–48. http:// teaching.ust.hk/~bee/papers/040918/1990-Kahneman-endowment _effect_coase_theorem.pdf.

Kaplan, Thomas. "Homeowners in Flood Zones Opt to Rebuild, Not Move." *New York Times* (April 27, 2013): A17. http://www.nytimes .com/2013/04/27/nyregion/new-yorks-storm-recovery-plan-gets -federal-approval.html.

Karplus, Valerie J., Sergey Paltsev, Mustafa Babiker, and John M. Reilly. "Should a Vehicle Fuel Economy Standard Be Combined with an Economy-Wide Greenhouse Gas Emissions Constraint? Implications for Energy and Climate Policy in the United States." *Energy Economics* 36 (2013): 322–33. http://www.sciencedirect.com/science/article/pii /S0140988312002150.

Keeling, C. D., S. C. Piper, R. B. Bacastow, M. Wahlen, T. P. Whorf, M. Heimann, and H. J. Meijer. *Exchanges of Atmospheric CO_2 and $13CO_2$ with the Terrestrial Biosphere and Oceans from 1978 to 2013. Global Aspects*. SIO Reference Series, No. 01–06, Scripps Institution of Oceanography, San Diego (accessed 2013). http://scrippsco2.ucsd.edu/data /in_situ_co2/monthly_mlo.csv.

Keith, David W. *A Case for Climate Engineering*. A Boston Review Book / MIT Press, 2013. http://mitpress.mit.edu/books/case-climate -engineering.

———. "Geoengineering the Climate: History and Prospect." *Annual Review of Energy and the Environment* 25.1 (2000): 245–84. http:// www.annualreviews.org/doi/abs/10.1146/annurev.energy.25.1.245 ?journalCode=energy.2.

———. "Photophoretic Levitation of Engineered Aerosols for Geoengineering." *Proceedings of the National Academy of Sciences* 107.38 (2010): 16428–31. http://www.pnas.org/content/107/38/16428.full.

Kemp, A. C., and B. P. Horton. "Contribution of Relative Sea-Level Rise to Historical Hurricane Flooding in New York City." *Journal of Quaternary Science* 28.6 (2013): 537–41. http://onlinelibrary.wiley.com/doi /10.1002/jqs.2653/abstract.

Keohane, Nathaniel O. "Cap and Trade, Rehabilitated: Using Tradable Permits to Control U.S. Greenhouse Gases." *Review of Environmental Economics and Policy* 3.1 (2009): 42–62. http://reep.oxfordjournals .org/content/3/1/42.abstract.

Keohane, Nathaniel O., and Gernot Wagner. "Judge a Carbon Market by Its Cap, Not Its Prices." *Financial Times* (July 17, 2013). http://www .ft.com/cms/s/0/de783c62-ee23-11e2-816e-00144feabdc0.html.

Khatiwala, S., F. Primeau, and T. Hall. "Reconstruction of the History of Anthropogenic CO_2 Concentrations in the Ocean." *Nature* 462.7271 (2009): 346–49. http://www.nature.com/nature/journal/v462/n7271 /full/nature08526.html.

Kirk-Davidoff, Daniel B., Eric J. Hintsa, James G. Anderson, and David W. Keith. "The Effect of Climate Change on Ozone Depletion through Changes in Stratospheric Water Vapour." *Nature* 402.6760 (1999): 399–401. http://www.nature.com/nature/journal/v402/n6760 /abs/402399a0.html.

Kirschbaum, Erik. "Germany Sets New Solar Power Record, Institute Says." Reuters (May 2012). http://www.reuters.com/article/2012/05/26 /us-climate-germany-solaridUSBRE84P0FI20120526.

Klepper, Gernot, and Wilfried Rickels. "The Real Economics of Climate Engineering." *Economics Research International* 2012.316564 (2012). http://www.hindawi.com/journals/econ/2012/316564/.

Klein, Naomi. "Capitalism vs. the Climate." *Nation* (November 28, 2011). http://www.thenation.com/article/164497/capitalism-vs-climate.

———. *This Changes Everything: Capitalism vs. the Climate*. Penguin, 2014.

Klier, Thomas, and Joshua Linn. "New-Vehicle Characteristics and the Cost of the Corporate Average Fuel Economy Standard." *RAND Journal of Economics* 43.1 (2012): 186–213. http://www.rff.org/rff /Documents/RFF-DP-10-50.pdf.

Klingman, William K., and Klingman, Nicholas P. *The Year without a Summer: 1816 and the Volcano That Darkened the World and Changed History*. St. Martin's Press, 2013.

Knight, Frank H. *Risk, Uncertainty, and Profit*. Hart, Schaffner and Marx, 1921. http://www.econlib.org/library/Knight/knRUP.html.

Knutti, Reto, and Gabriele C. Hegerl. "The Equilibrium Sensitivity of the Earth's Temperature to Radiation Changes." *Nature Geoscience* 1.11 (2008): 735–43. http://www.nature.com/ngeo/journal/v1/n11/abs /ngeo337.html.

Kolbert, Elizabeth. *Field Notes from a Catastrophe: Man, Nature, and Climate Change*. Bloomsbury, 2006. http://books.google.com/books ?id=Bd-uEKO7g4oC.

———. "Hosed." *New Yorker* (November 16, 2009). http://www.new yorker.com/arts/critics/books/2009/11/16/091116crbo_books_kolbert.

———. *The Sixth Extinction: An Unnatural History*. Henry Holt, 2014.

Kopp, Robert E., and Bryan K. Mignone. "The U.S. Government's Social Cost of Carbon Estimates after Their First Two Years: Pathways for Improvement." *Economics: The Open-Access, Open-Assessment E-Journal* 6 (2012–15): 1–41. http://dx.doi.org/10.5018/economics-ejournal.ja .2012-15.

Kravitz, Ben, Alan Robock, Luke Oman, Georgiy Stenchikov, and Allison B. Marquardt. "Sulfuric Acid Deposition from Stratospheric Geoengineering with Sulfate Aerosols." *Journal of Geophysical Research: Atmospheres (1984–2012)* 114.D14 (2009). http://onlinelibrary.wiley .com/doi/10.1029/2009JD011918/abstract.

Krosnick, Jon A. "The Climate Majority." *New York Times* (June 8, 2010). http://www.nytimes.com/2010/06/09/opinion/09krosnick.html ?pagewanted=all&_r=1&.

"Kyoto Protocol to the united nations framework convention on climate change." *United Nations Framework Convention on Climate Change* (1997). http://unfccc.int.

Laibson, David. "Golden Eggs and Hyperbolic Discounting." *Quarterly Journal of Economics* 112.2 (1997): 443–77. http://scholar.harvard.edu/ laibson/publications/golden-eggs-and-hyperbolic-discounting.

Latham, John, Keith Bower, Tom Choularton, Hugh Coe, Paul Connolly, Gary Cooper, Tim Craft et al. "Marine Cloud Brightening." *Philosophical Transactions of the Royal Society: Mathematical, Physical and Engineering Sciences* 370.1974 (2012): 4217–62. http://rsta.royal societypublishing.org/content/370/1974/4217.short.

Latham, John, Philip Rasch, Chih-Chieh Chen, Laura Kettles, Alan Gadian, Andrew Gettelman, Hugh Morrison, Keith Bower, and Tom Choularton. "Global Temperature Stabilization via Controlled Albedo Enhancement of Low-Level Maritime Clouds." *Philosophical Transactions of the Royal Society* 366.1882 (2008): 3969–87. http://rsta .royalsocietypublishing.org/content/366/1882/3969.short.

Lenton, Timothy M., Hermann Held, Elmar Kriegler, Jim W. Hall, Wolfgang Lucht, Stefan Rahmstorf, and Hans Joachim Schellnhuber. "Tipping Elements in the Earth's Climate System." *Proceedings of the National Academy of Sciences* 105.6 (2008): 1786–93. http://www.pnas .org/content/105/6/1786.full.pdf.

Le Quéré, C., G. P. Peters, R. J. Andres, R. M. Andrew, T. Boden, P. Ciais, P. Friedlingstein et al. "Global Carbon Budget 2013." *Earth System Science Data Discussions* 6.1 (2014): 235–63. http://www.earth-syst-sci -data.net/6/235/2014/essd-6-235-2014.html.

Leurig, Sharlene, and Andrew Dlugolecki. *Insurer Climate Risk Disclosure Survey: 2012 Findings and Recommendations.* Ceres (March 2013). http://www.ceres.org/resources/reports/naic-report.

Levitt, Steven, and Stephen Dubner. *SuperFreakonomics*. Harper Collins, 2010. http://www.superfreakonomicsbook.com/.

Lewis, Robert and Al Shaw. "After Sandy, Government Lends to Rebuild in Flood Zones." *ProPublica* and WNYC (March 2013). http:// www.propublica.org/article/after-sandy-government-lends-to-rebuild -in-flood-zones.

Li, Qingxiang, Jiayou Huang, Zhihong Jiang, Liming Zhou, Peng Chu, and Kaixi Hu. "Detection of Urbanization Signals in Extreme Winter Minimum Temperature Changes over Northern China." *Climatic Change* 122.4 (2014): 595–608. http://link.springer.com/article /10.1007/s10584-013-1013-z.

Liebreich, Michael. "Global Trends in Clean Energy Investment." *Bloomberg New Energy Finance.* Presentation given at the Clean Energy Ministerial in Delhi (April 17, 2013). http://about.bnef.com /presentations/global-trends-in-clean-energy-investment/.

Liger-Belair, Gérard, Marielle Bourget, Sandra Villaume, Philippe Jeandet, Hervé Pron, and Guillaume Polidori. "On the Losses of Dissolved CO_2 during Champagne Serving." *Journal of Agricultural and Food Chemistry* 58.15 (2010): 8768–75. http://pubs.acs.org/doi/abs/10.1021/jf101239w.

Liger-Belair, Gérard, Guillaume Polidori, and Philippe Jeandet. "Recent Advances in the Science of Champagne Bubbles." *Chemical Society Reviews* 37.11 (2008): 2490–511. http://www.ncbi.nlm.nih.gov/pub med/18949122.

Lin, Ning, Kerry Emanuel, Michael Oppenheimer, and Erik Vanmarcke. "Physically Based Assessment of Hurricane Surge Threat under Climate Change." *Nature Climate Change* 2 (2012): 462–67. http://www .nature.com/nclimate/journal/v2/n6/abs/nclimate1389.html.

Litterman, Robert B. "The Other Reason for Divestment." Ensia .com (November 5, 2013). http://ensia.com/voices/the-other-reason -for-divestment/.

———. "What Is the Right Price for Carbon Emissions?" *Regulation* (Summer 2013). http://www.cato.org/sites/cato.org/files/serials/files /regulation/2013/6/regulation-v36n2-1-1.pdf.

Lott, Fraser C., Nikolaos Christidis, and Peter A. Stott. "Can the 2011 East African Drought Be Attributed to Human-Induced Climate Change?" *Geophysical Research Letters* 40.6 (2013): 1177–81. http:// onlinelibrary.wiley.com/doi/10.1002/grl.50235/abstract.

Lovett, Ken. "Gov. Cuomo: Sandy as Bad as Anything I've Experienced in New York." *New York Daily News* (October 30, 2012). http:// www.nydailynews.com/blogs/dailypolitics/2012/10/gov-cuomo -sandy-as-bad-as-anything-ive-experienced-in-new-york.

Lynas, Mark. *Six Degrees: Our Future on a Hotter Planet.* Fourth Estate, 2007.

Machiavelli, Niccolò. *The Prince.* 1532. Chapter 11. Translated by W. K. Marriott in 1908. http://www.constitution.org/mac/prince06.htm.

MacKay, David J.C. *Sustainable Energy—without the Hot Air.* UIT Cambridge, 2009. www.withouthotair.com.

Major, Julie, Johannes Lehmann, Marco Rondon, and Christine Goodale. "Fate of Soil-Applied Black Carbon: Downward Migration, Leaching and Soil Respiration." *Global Change Biology* 16 (2010): 1366–79. http://www.css.cornell.edu/faculty/lehmann/publ/Global ChangeBiol%2016,%201366-1379,%202010%20Major.pdf.

Margolis, Joshua D., Hillary Anger Elfenbein, and James P. Walsh. "Does It Pay to Be Good . . . and Does It Matter? A Meta-Analysis of the Relationship between Corporate Social and Financial Performance." SSRN Working paper (March 1, 2009). http://ssrn.com /abstract=1866371.

Marlon, J.R., Leiserowitz, A., and Feinberg, G. "Scientific and Public Perspectives on Climate Change." Yale University, Yale Project on Climate Change Communication (2013). http://environment.yale .edu/climate-communication/files/ClimateNote_Consensus_Gap _May2013_FINAL6.pdf.

Mastrandrea, Michael D., Katharine J. Mach, Gian-Kasper Plattner, Ottmar Edenhofer, Thomas F. Stocker, Christopher B. Field, Kristie L. Ebi, and Patrick R. Matschoss. "The IPCC AR5 Guidance Note on Consistent Treatment of Uncertainties: A Common Approach across the Working Groups." *Climatic Change* 108.4 (2011): 675–91. http://link.springer.com/content/pdf/10.1007/s10584-011-0178-6 .pdf.

Matthews, H. Damon, and Ken Caldeira. "Transient Climate–Carbon Simulations of Planetary Geoengineering." *Proceedings of the National Academy of Sciences* 104.24 (2007): 9949–54. http://www.pnas.org /content/104/24/9949.short.

Matthews, H. Damon, Nathan P. Gillett, Peter A. Stott, and Kirsten Zickfeld. "The Proportionality of Global Warming to Cumulative Carbon Emissions." *Nature* 459.7248 (2009): 829–32. http://www.nature.com /nature/journal/v459/n7248/full/nature08047.html.

Mauna Loa Observatory, National Oceanic and Atmospheric Administration (NOAA). http://www.esrl.noaa.gov/gmd/obop/mlo/.

McClellan, Justin, David W. Keith, and Jay Apt. "Cost Analysis of Stratospheric Albedo Modification Delivery Systems." *Environmental Research Letters* 7.3 (2012): 034019. http://iopscience.iop.org/1748 -9326/7/3/034019.

McCormick, M. Patrick, Larry W. Thomason, and Charles R. Trepte. "Atmospheric Effects of the Mt. Pinatubo Eruption." *Nature* 373.6513 (1995): 399–404. http://www.nature.com/nature/journal/v373/n6513 /abs/373399a0.html.

McGranahan, Gordon, Deborah Balk, and Bridget Anderson. "The Rising Tide: Assessing the Risks of Climate Change and Human Settlements in Low Elevation Coastal Zones." *Environment and Urbanization* 19.1 (2007): 17–37. http://eau.sagepub.com/content/19/1/17.

McKibben, Bill. "Global Warming's Terrifying New Math." *Rolling Stone* (August 2, 2012). http://www.rollingstone.com/politics/news /global-warmings-terrifying-new-math-20120719.

Meehl, Gerald A., Aixue Hu, Claudia Tebaldi, Julie M. Arblaster, Warren M. Washington, Haiyan Tang, Benjamin M. Sanderson, Toby Ault, Warren G. Strand, and James B. White III. "Relative Outcomes of Climate Change Mitigation Related to Global Temperature versus Sea-Level Rise." *Nature Climate Change* 2 (2012): 576–80. http://www.nature.com/nclimate/journal/v2/n8/full/nclimate1529 .html.

Mehra, Rajnish. "The Equity Premium Puzzle: A Review." *Foundations and Trends in Finance*, 2.1 (2008): 1–81. http://papers.ssrn.com/sol3 /papers.cfm?abstract_id=1624986.

Melillo, Jerry M., Terese (T.C.) Richmond, and Gary W. Yohe, eds.: *Climate Change Impacts in the United States: The Third National Climate Assessment*. U.S. Global Change Research Program (2014).

Mellström, Carl, and Magnus Johannesson. "Crowding Out in Blood Donation: Was Titmuss Right?" *Journal of the European Economic Association* 6.4 (2008): 845–63. http://onlinelibrary.wiley.com/doi/10.1162 /JEEA.2008.6.4.845/abstract.

Mendelsohn, Robert, Kerry Emanuel, Shun Chonabayashi, and Laura Bakkensen. "The Impact of Climate Change on Global Tropical Cyclone Damage." *Nature Climate Change* 2.3 (2012): 205–9. http://www .nature.com/nclimate/journal/v2/n3/full/nclimate1357.html.

Mendelsohn, Robert, Wendy Morrison, Michael E. Schlesinger, and Natalia G. Andronova. "Country-Specific Market Impacts of Climate Change." *Climatic Change* 45.3–4 (2000): 553–69. http://link.springer .com/article/10.1023%2FA%3A1005598717174.

Meng, Kyle C. "Estimating the Cost of Climate Policy Using Prediction Markets and Lobbying Records." Ph.D. dissertation, Columbia University, 2013. https://dl.dropboxusercontent.com/u/3015077/Website /Meng_cap_trade_Aug2013.pdf.

Menon, Surabi, Hashem Akbari, Sarith Mahanama, Igor Sednev, and Ronnen Levinson. "Radiative Forcing and Temperature Response to Changes in Urban Albedos and Associated CO_2 Offsets," *Environmental Research Letters* 5.1 (2010). http://iopscience.iop.org/1748 -9326/5/1/014005.

Metcalf, Gilbert E., "Designing a Carbon Tax to Reduce U.S. Greenhouse Gas Emissions." *Review of Environmental Economics and Policy* 3.1 (2009): 63–83. http://reep.oxfordjournals.org/content/3/1/63.abstract.

Metz, Tim, Alan Murray, Thomas E. Ricks, and Beatrice E. Garcia. "The Crash of '87: Stocks Plummet 508 Amid Panicky Selling." *Wall Street Journal* (October 20, 1987). http://online.wsj.com/article/SB10000872 3963904447348045780645715935981 96.html.

Miller, Bruce G. *Coal Energy Systems*. Elsevier Academic Press, 2005. http://books.google.com/books/about/Coal_Energy_Systems.html ?id=PYyJEEyJN94C.

Millner, Antony, Simon Dietz, and Geoffrey Heal. "Scientific Ambiguity and Climate Policy." *Environmental and Resource Economics* 55.1 (2013): 21–46.

Monastersky, R. "Global Carbon Dioxide Levels Near Worrisome Milestone." *Nature* 497.7447 (2013): 13. http://www.nature.com/news /global-carbon-dioxide-levels-near-worrisome-milestone-1.12900.

Morris, Eric. "From Horse Power to Horsepower." *Access* 30 (2007): 2–9. http://www.uctc.net/access/30/Access%2030%20-%2002%20-%20 Horse%20Power.pdf.

Morris, Errol. "The Certainty of Donald Rumsfeld." *New York Times* Opinionator, 4-part series (March 2014). http://opinionator.blogs .nytimes.com/2014/03/25/the-certainty-of-donald-rumsfeld -part-1/.

Morris, Ian. *Why the West Rules—for Now: The Patterns of History and What They Reveal about the Future.* Farrar Straus and Giroux, 2010. http://books.google.com/books?id=qNVrfoSubmIC.

Moyer, Elisabeth, Michael D. Woolley, Michael Glotter, and David A. Weisbach. "Climate Impacts on Economic Growth as Drivers of Uncertainty in the Social Cost of Carbon." *Center for Robust Decision Making on Climate and Energy Policy Working Paper* 13 (2013): 25. http://www .law.uchicago.edu/files/file/652-ejm-mdw-mjg-daw-climate-new .pdf.

Nabuurs, Gert-Jan, Marcus Lindner, Pieter J. Verkerk, Katja Gunia, Paola Deda, Roman Michalak, and Giacomo Grassi. "First Signs of Carbon Sink Saturation in European Forest Biomass." *Nature Climate Change* 3.9 (2013): 792–96. http://www.nature.com/nclimate/journal/v3/n9 /full/nclimate1853.html.

"NASA Authorization Act of 2005" (Public Law No. 109–155, 119 Stat. 2895, 2005). http://www.gpo.gov/fdsys/pkg/PLAW-109publ155/pdf /PLAW-109publ155.pdf.

National Disaster Risk Reduction and Management Council. Situation Report No. 38 re Effects of Typhoon "PABLO" (BOPHA), December 25, 2012. http://www.ndrrmc.gov.ph/attachments/article/835 /Update%20Sitrep%20No.%2038.pdf.

"Near-Earth Object Survey and Deflection Analysis of Alternatives." NASA, 2007. http://www.nasa.gov/pdf/171331main_NEO_report _march07.pdf.

Newell, Richard G., and William A. Pizer. "Regulating Stock Externalities under Uncertainty." *Journal of Environmental Economics and Management* 45.2 (2003): 416–32. http://www.sciencedirect.com/science /article/pii/S0095069602000165.

Nguyen B. T., J. Lehmann, J. Kinyangi, R. Smernik, S. J. Riha, and M. H. Engelhard. "Long-Term Black Carbon Dynamics in Cultivated Soil." *Biogeochemistry* 89.3 (2008): 295–308. http://link.springer.com/article /10.1007%2Fs10533-008-9220-9.

"Nigeria Restores Fuel Subsidy to Quell Nationwide Protests." *Guardian* (January 16, 2012). http://www.guardian.co.uk/world/2012/jan/16 /nigeria-restores-fuel-subsidy-protests.

Nordhaus, William D. *The Climate Casino: Risk, Uncertainty, and Economics for a Warming World.* Yale University Press, 2013. http:// www.amazon.com/The-Climate-Casino-Uncertainty-Economics /dp/030018977X.

Nordhaus, William D. "Economic Aspects of Global Warming in a Post-Copenhagen Environment." *Proceedings of the National Academy of Sciences* 107.26 (2010): 11721–26. http://www.pnas.org/content/107/26/11721.full.

———. "Estimates of the Social Cost of Carbon: Concepts and Results from the DICE-2013R Model and Alternative Approaches." *Journal of the Association of Environmental and Resource Economists* 1.1 (2014): 273–312. http://dx.doi.org/10.1086/676035.

———. "Optimal Greenhouse Gas Reductions and Tax Policy in the 'DICE' Model." *American Economic Review* 83.2 (1993): 313–17. http://ideas.repec.org/a/aea/aecrev/v83y1993i2p313-17.html.

———. "An Optimal Transition Path for Controlling Greenhouse Gases." *Science* 258.5086 (1992): 1315–19. http://www.sciencemag.org/content/258/5086/1315.

———. "To Slow or Not to Slow: The Economics of the Greenhouse Effect." *Economic Journal* 101.407 (1991): 920–37. http://ideas.repec.org/a/ecj/econjl/v101y1991i407p920-37.html.

Normile, Dennis. "Clues to Supertyphoon's Ferocity Found in the Western Pacific." *Science* 342.6162 (2013): 1027. http://www.sciencemag.org/content/342/6162/1027.short.

OECD. *Inventory of Estimated Budgetary Support and Tax Expenditures for Fossil Fuels 2013*. OECD Publishing, 2013. http://dx/doi.org/10.1787/9789264187610-en.

Office of Management and Budget (OMB). "Circular No. A-94 Revised" (October 29, 1992). http://www.whitehouse.gov/omb/circulars_a094.

Ogburn, Stephanie Paige. "How Media Pushed Climate Change 'Pause' into the Mainstream." *Environment and Energy Publishing* (November 4, 2013). http://www.eenews.net/stories/1059989863.

———. "What's in a Name? Would 'the Pause' by Any Other Name Help Scientists Communicate?" *Environment and Energy Publishing* (November 1, 2013). http://www.eenews.net/stories/1059989790.

Oleson, Keith W., G. B. Bonan, and J. Feddema. "Effects of White Roofs on Urban Temperature in a Global Climate Model." *Geophysical Research Letters* 37.3 (2010). http://onlinelibrary.wiley.com/doi/10.1029/2009GL042194/abstract.

Olivier, Jos, Greet Janssens-Maenhous, Marilena Muntean, and Jeroen Peters. "Trends in Global CO_2 Emissions: 2013 Report." PBL Netherlands Environmental Assessment Agency and European Commission Joint Research Center, 2013. http://edgar.jrc.ec.europa.eu/news_docs/pbl-2013-trends-in-global-co2-emissions-2013-report-1148.pdf.

"Operation Ivy." U.S. Nuclear Weapons Archive (last modified 1999). http://nuclearweaponarchive.org/Usa/Tests/Ivy.html.

Otto, F.E.L., N. Massey, G. J. Oldenborgh, R. G. Jones, and M. R. Allen. "Reconciling Two Approaches to Attribution of the 2010 Russian Heat Wave." *Geophysical Research Letters* 39.4 (2012). http://online library.wiley.com/doi/10.1029/2011GL050422/abstract.

Pall, Pardeep, Tolu Aina, Daithi A. Stone, Peter A. Stott, Toru Nozawa, Arno G. J. Hilberts, Dag Lohmann, and Myles R. Allen. "Anthropogenic Greenhouse Gas Contribution to Flood Risk in England and Wales in Autumn 2000." *Nature* 470.7334 (2011): 382–85. http://www .nature.com/nature/journal/v470/n7334/abs/nature09762.html.

Palumbi, Stephen R., Daniel J. Barshis, Nikki Traylor-Knowles, and Rachael A. Bay. "Mechanisms of Reef Coral Resistance to Future Climate Change." *Science* 344.6186 (2014): 895–98. http://www .sciencemag.org/content/early/2014/04/23/science.1251336 .abstract.

Parfit, Derek "Five Mistakes in Moral Mathematics." Chap. 3 of *Reasons and Persons*. Oxford University Press, 1986. 67–86. http://www .oxfordscholarship.com/view/10.1093/019824908X.001.0001/acprof -9780198249085.

Parris, A., P. Bromirski, V. Burkett, D. Cayan, M. Culver, J. Hall, R. Horton, K. Knuuti, R. Moss, J. Obeysekera, A. Sallenger, and J. Weiss. "Global Sea Level Rise Scenarios for the United States National Climate Assessment." National Oceanic and Atmospheric Administration Tech Memo OAR CPO (2012). http://cpo.noaa.gov/sites/cpo /Reports/2012/NOAA_SLR_r3.pdf.

Parson, Edward A. "The Big One: A Review of Richard Posner's Catastrophe: Risk and Response." *Journal of Economic Literature* 45.1 (2007): 147–64. http://www.jstor.org/discover/10.2307/27646750.

Parson, Edward A., and David W. Keith. "End the Deadlock on Governance of Geoengineering Governance." *Science* 339.6125 (2013): 1278–79. http://www.sciencemag.org/content/339/6125/1278.

Peterson, Thomas C., Peter A. Stott, and Stephanie Herring. "Explaining Extreme Events of 2011 from a Climate Perspective." *Bulletin of the American Meteorological Society* 93.7 (2012): 1041–67. http://journals .ametsoc.org/doi/abs/10.1175/BAMS-D-12-00021.1.

Pew Global Attitudes Project (June 2013). http://www.pewglobal.org /files/2013/06/Pew-Research-Center-Global-Attitudes-Project-Global -Threats-Report-FINAL-June-24-20131.pdf.

Pew Research Center / USA Today Survey (February 21 2013). http:// www.people-press.org/files/legacy-pdf/02-21-13%20Political%20 Release.pdf.

"Philippines: Typhoon Haiyan Situation Report No. 34." United Nations Office for the Coordination of Humanitarian Affairs (2013).

http://reliefweb.int/sites/reliefweb.int/files/resources/OCHA PhilippinesTyphoonHaiyanSitrepNo.34.28Jan2014.pdf.

Pigou, Arthur. 1920. *The Economics of Welfare*. 4th ed. Macmillan, 1932. http://www.econlib.org/library/NPDBooks/Pigou/pgEW20.html #Part II, Chapter 9.

Piketty, Thomas. *Capital in the 21st Century*. Harvard University Press, 2014. http://www.hup.harvard.edu/catalog.php?isbn=9780674430006.

Pindyck, Robert S., "Climate Change Policy: What Do the Models Tell Us?" *Journal of Economic Literature* 51.3 (2013): 860–72. http://www.aeaweb.org/articles.php?f=s&doi=10.1257/jel.51.3.860.

Pongratz, Julia, D. B. Lobell, L. Cao, and K. Caldeira. "Crop Yields in a Geoengineered Climate." *Nature Climate Change* 2.2 (2012): 101–5. http://www.nature.com/nclimate/journal/v2/n2/full/nclimate1373.html.

Posner, Richard A. *Catastrophe: Risk and Response*. Oxford University Press, 2004. http://books.google.com/books/about/Catastrophe_Risk_and_Response.html?id=SDe59lXSrY8C.

"Pressure Continues: Stocks Sink Lower under Record Volume of Liquidation." *Wall Street Journal* (October 30, 1929).

Previdi, M., B. G. Liepert, D. Peteet, J. Hansen, D. J. Beerling, A. J. Broccoli, S. Frolking et al. "Climate Sensitivity in the Anthropocene." *Quarterly Journal of the Royal Meteorological Society* 139.674 (2013): 1121–31. http://onlinelibrary.wiley.com/doi/10.1002/qj.2165/abstract.

Pun, Iam-Fei, I-I. Lin, and Min-Hui Lo. "Recent Increase in High Tropical Cyclone Heat Potential Area in the Western North Pacific Ocean." *Geophysical Research Letters* 40.17 (2013): 4680–84. http://onlinelibrary.wiley.com/doi/10.1002/grl.50548/abstract.

Rahmstorf, Stefan, and Dim Coumou. "Increase of Extreme Events in a Warming World." *Proceedings of the National Academy of Sciences* 108.44 (2011): 17905–9. http://www.pnas.org/content/108/44/17905.short.

Ramanathan, Veerabhadran, and Yan Feng. "On Avoiding Dangerous Anthropogenic Interference with the Climate System: Formidable Challenges Ahead." *Proceedings of the National Academy of Sciences* 105.38 (2008): 14245–50. http://www.pnas.org/content/105/38/14245.short.

Rau, Greg H. "CO_2 Mitigation via Capture and Chemical Conversion in Seawater." *Environmental Science and Technology* 45.3 (2010): 1088–92. http://pubs.acs.org/doi/abs/10.1021/es102671x.

Rayner, Steve, Clare Heyward, Tim Kruger, Nick Pidgeon, Catherine Redgwell, and Julian Savulescu. "The Oxford Principles." *Climatic Change* 121.3 (2013): 499–512. http://link.springer.com/article/10.1007/s10584-012-0675-2.

Reichstein, Markus, Michael Bahn, Philippe Ciais, Dorothea Frank, Miguel D. Mahecha, Sonia I. Seneviratne, Jakob Zscheischier et al. "Climate Extremes and the Carbon Cycle." *Nature* 500.7462 (2013): 287–95. http://www.nature.com/nature/journal/v500/n7462/full/nature12350.html.

"Report: The After Action Review / Lessons Learned Workshops for Typhoon Bopha Response." United Nations Office for the Coordination of Humanitarian Affairs (2013). http://reliefweb.int/sites/reliefweb.int/files/resources/Bopha%20AAR-LLR%20Report%202013_FINAL_14%20June%202013.pdf.

Revesz, Richard, and Michael Livermore. *Retaking Rationality: How Cost-Benefit Analysis Can Better Protect the Environment and Our Health.* Oxford University Press, 2008. http://www.amazon.com/Retaking-Rationality-Benefit-Analysis-Environment/dp/0195368576.

Ricke, Katharine L., M. Granger Morgan, and Myles R. Allen. "Regional Climate Response to Solar-Radiation Management." *Nature Geoscience* 3.8 (2010): 537–41. http://www.nature.com/ngeo/journal/v3/n8/full/ngeo915.html.

Rignot, E., J. Mouginot, M. Morlighem, H. Seroussi, and B. Scheuchl. "Widespread, Rapid Grounding Line Retreat of Pine Island, Thwaites, Smith and Kohler Glaciers, West Antarctica from 1992 to 2011." *Geophysical Research Letters* 41.10 (2014): 3502–9. http://onlinelibrary.wiley.com/doi/10.1002/2014GL060140/abstract.

Risky Business Project. *Risky Business: The Economic Risks of Climate Change in the United States.* 2014. http://riskybusiness.org.

Robine, Jean-Marie, Siu Lan K. Cheung, Sophie Le Roy, Herman Van Oyen, Claire Griffiths, Jean-Pierre Michel, François Richard Herrmann. "Death Toll Exceeded 70,000 in Europe during the Summer of 2003." *Comptes rendus biologies* 331.2 (2008): 171–78. http://www.sciencedirect.com/science/article/pii/S1631069107003770.

Robock, Alan. "20 Reasons Why Geoengineering May Be a Bad Idea." *Bulletin of the Atomic Scientists* 64.2 (2008): 14–18, 59. http://www.atmos.washington.edu/academics/classes/2012Q1/111/20Reasons.pdf.

———. "Is Geoengineering Research Ethical?" *Peace and Security* 4 (2012): 226–29. http://climate.envsci.rutgers.edu/pdf/GeoResearchEthics.pdf.

Robock, Alan, L. Oman, and G. L. Stenchikov. "Regional Climate Responses to Geoengineering with Tropical and Arctic SO_2 Injections." *Journal of Geophysical Research* 113.D16 (2008): D16101. http://onlinelibrary.wiley.com/doi/10.1029/2008JD010050/abstract.

Roe, Gerard. "Costing the Earth: A Numbers Game or a Moral Imperative?" *Weather, Climate, and Society* 5.4 (2013): 378–80. http://journals.ametsoc.org/doi/abs/10.1175/WCAS-D-12-00047.1.

Roe, Gerard H., and Yoram Bauman. "Climate Sensitivity: Should the Climate Tail Wag the Policy Dog?" *Climatic Change* 117.4 (2013): 647–62. http://link.springer.com/article/10.1007/s10584-012-0582-6.

Rosenthal, Elisabeth. "Your Biggest Carbon Sin May Be Air Travel." *New York Times* (January 27, 2013): SR4. http://www.nytimes.com /2013/01/27/sunday-review/the-biggest-carbon-sin-air-travel.html.

Rosenzweig, C. and W. Solecki, eds. "Climate Risk Information 2013: Observations, Climate Change Projections, and Maps." New York City Panel on Climate Change (June 2013). http://www.nyc.gov/html/planyc2030 /downloads/pdf/npcc_climate_risk_information_2013_report.pdf.

Roston, Eric. *The Carbon Age: How Life's Core Element Has Become Civilization's Greatest Threat*. Bloomsbury Publishing USA, 2009. http:// www.ericroston.com/.

Rottenberg, Dan. *In the Kingdom of Coal*. Routledge, 2003. http://books .google.com/books/about/In_the_Kingdom_of_Coal.html?id=VL 8YWx2X8asC.

Rowley, Rex J., John C. Kostelnick, David Braaten, Xingong Li, and Joshua Meisel. "Risk of Rising Sea Level to Population and Land Area." *Eos, Transactions American Geophysical Union* 88.9 (2007): 105–7. http://onlinelibrary.wiley.com/doi/10.1029/2007EO090001/abstract.

Royal Society. "Geoengineering the Climate: Science, Governance and Uncertainty" (September 2009). http://royalsociety.org/uploaded Files/Royal_Society_Content/policy/publications/2009/8693.pdf.

Rumsfeld, Donald. "Press Conference at NATO Headquarters." Brussels, Belgium (June 6, 2002). http://www.defense.gov/transcripts/transcript .aspx?transcriptid=3490.

Rybczynski, Natalia, John C. Gosse, C. Richard Harington, Roy A. Wogelius, Alan J. Hidy, and Mike Buckley. "Mid-Pliocene Warm-Period Deposits in the High Arctic Yield Insight into Camel Evolution." *Nature Communications* 4 (2013): 1550. http://www.nature.com/ncomms /journal/v4/n3/full/ncomms2516.html.

Salter, Stephen, Graham Sortino, and John Latham, "Sea-Going Hardware for the Cloud Albedo Method of Reversing Global Warming." *Philosophical Transactions of the Royal Society* 366.1882 (2008): 3989–4006. http://rsta.royalsocietypublishing.org/content/366/1882/3989.full.

Samuelson, William, and Richard Zeckhauser. "Status Quo Bias in Decision Making." *Journal of Risk and Uncertainty* 1.1 (1988): 7–59. http://dtserv2.compsy.uni-jena.de/__C125757B00364C53.nsf/0 /F0CC3CAE039C8B42C125757B00473C77/$FILE/samuelson_zeck hauser_1988.pdf.

Sandel, Michael J. *Justice: What's the Right Thing to Do?* Farrar Strauss and Giroux, 2009. http://www.justiceharvard.org/.

————. "Market Reasoning as Moral Reasoning: Why Economists Should Re-engage with Political Philosophy." *Journal of Economic Perspectives* 27.4 (2013): 121–40. http://pubs.aeaweb.org/doi/pdfplus /10.1257/jep.27.4.121.

Sandsmark, Maria, and Haakon Vennemo, "A Portfolio Approach to Climate Investments: CAPM and Endogenous Risk." *Environmental and Resource Economics* 37.4 (2007): 681–95. http://link.springer.com /article/10.1007/s10640-006-9049-4.

Schelling, Thomas, "The Economic Diplomacy of Geoengineering." *Climatic Change* 33.3 (1996): 303–7. http://link.springer.com /article/10.1007%2FBF00142578.

Schlosser, Eric. *Command and Control.* Penguin, 2013.

Schmidt, Gavin, and Stefan Rahmstorf. "11°C Warming, Climate Crisis in 10 Years?" RealClimate.org (January 29, 2005). http://www.real climate.org/index.php/archives/2005/01/climatepredictionnet -climate-challenges-and-climate-sensitivity/.

Schneider, Stephen H. *Science as a Contact Sport.* National Geographic Society, 2009. http://www.amazon.com/Science-Contact-Sport-Inside -Climate/dp/1426205406.

Self, Stephen, Jing-Xia Zhao, Rick E. Holasek, Ronnie C. Torres, and Alan J. King. "The Atmospheric Impact of the 1991 Mount Pinatubo Eruption." U.S. Geological Survey, 1999. http://pubs.usgs.gov /pinatubo/self/index.html.

Shepherd, Andrew, Erik R. Ivins, A. Geruo, Valentina R. Barletta, Mike J. Bentley, Srinivas Bettadpur, Kate H. Briggs et al. "A Reconciled Estimate of Ice-Sheet Mass Balance." *Science* 338.6111 (2012): 1183–89. http://www.sciencemag.org/content/338/6111/1183.

Sherwood, Steven C., Sandrine Bony, and Jean-Louis Dufresne. "Spread in Model Climate Sensitivity Traced to Atmospheric Convective Mixing." *Nature* 505.7481 (2014): 37–42. http://www.nature.com/nature /journal/v505/n7481/full/nature12829.html.

Shoemaker, Julie K., and Daniel P. Schrag. "The Danger of Overvaluing Methane's Influence on Future Climate Change." *Climatic Change* 120.4 (2013): 903–14. http://link.springer.com/article/10.1007/s10584 -013-0861-x.

Silver, Nate. "Crunching the Risk Numbers." *Wall Street Journal* (January 8, 2010). http://online.wsj.com/news/articles/SB100014240527487034 8100457464696371306511.

Socolow, Robert. "Truths We Must Tell Ourselves to Manage Climate Change." *Vanderbilt Law Review* 65.6 (2012): 1455–78. http://www .vanderbiltlawreview.org/content/articles/2012/11/Socolow_-65 _Vand_L_Rev_1455.pdf.

Solomon, Pierrehumbert, Damon Matthews, John S. Daniel, and Pierre Friedlingstein. "Atmospheric Composition, Irreversible Climate Change, and Mitigation Policy." In *Climate Science for Serving Society.* Springer Netherlands, 2013. 415–36. http://link.springer.com/chapter/10.1007/978-94-007-6692-1_15.

Solomon, Susan, Gian-Kasper Plattner, Reto Knutti, and Pierre Friedlingstein. "Irreversible Climate Change Due to Carbon Dioxide Emissions." *Proceedings of the National Academy of Sciences* 106.6 (2009): 1704–9. http://www.pnas.org/content/early/2009/01/28/0812721106.abstract.

Specter, Michael. "The First Geo-Vigilante." *New Yorker* (October 2012). http://www.newyorker.com/online/blogs/newsdesk/2012/10/the-first-geo-vigilante.html.

"Statistic Data on the German Solar Power (Photovoltaic) Industry." German Solar Industry Association (BSW-Solar) (June 2013). http://www.solarwirtschaft.de/fileadmin/media/pdf/2013_2_BSW-Solar_fact_sheet_solar_power.pdf.

Stenchikov, Georgiy L., Ingo Kirchner, Alan Robock, Hans-F. Graf, Juan Carlos Antuna, R. G. Grainger, Alyn Lambert, and Larry Thomason. "Radiative Forcing from the 1991 Mount Pinatubo Volcanic Eruption." *Journal of Geophysical Research: Atmospheres (1984–2012)* 103. D12 (1998): 13837–57. http://onlinelibrary.wiley.com/doi/10.1029/98JD00693/abstract.

Stephens, Phillip. "Major Puts ERM Membership on Indefinite Hold." *Financial Times* (September 25, 1992).

Sterman, John D. "Risk Communication on Climate: Mental Models and Mass Balance." *Science* 322.5901 (2008): 532–33. http://www.sciencemag.org/content/322/5901/532.summary.

Stern, Nicholas. "The Structure of Economic Modeling of the Potential Impacts of Climate Change: Grafting Gross Underestimation of Risk onto Already Narrow Science Models." *Journal of Economic Literature* 51.3 (2013): 838–59. http://www.aeaweb.org/articles.php?doi=10.1257/jel.51.3.838.

Sterner, Thomas, ed. *Fuel Taxes and the Poor: The Distributional Consequences of Gasoline Taxation and Their Implications for Climate Policy.* RFF Press, 2011. http://www.routledge.com/books/details/9781617260926/.

Sterner, Thomas, and U. Martin Persson. "An Even Sterner Review: Introducing Relative Prices into the Discounting Debate." *Review of Environmental Economics and Policy* 2.1 (2008): 61–76. http://reep.oxfordjournals.org/content/2/1/61.short.

Stocker, Thomas F. "The Closing Door of Climate Targets." *Science* 339.6117 (2013): 280–82. http://www.sciencemag.org/content/339/6117/280.full ?sid=a0ad9885-1715-4bff-a27f-54d5f68c15e2#ref-2.

Stommel, Henry M. *Volcano Weather: The Story of 1816, the Year without a Summer*. Seven Seas Press, 1983.

Stothers, Richard B. "The Great Tambora Eruption in 1815 and Its Aftermath." *Science* 224.4654 (1984): 1191–98. http://www.sciencemag.org /content/224/4654/1191.short.

Stott, Peter A., Myles Allen, Nikolaos Christidis, Randall M. Dole, Martin Hoerling, Chris Huntingford, Pardeep Pall, Judith Perlwitz, and Dáithí Stone. "Attribution of Weather and Climate-Related Events." In *Climate Science for Serving Society*, edited by Ghassem R. Asrar and James W. Hurrell. Netherlands: Springer, 2013. 307–37. http://link .springer.com/chapter/10.1007%2F978-94-007-6692-1_12.

Stott, Peter A., Dáithí Stone, and Myles Allen. "Human Contribution to the European Heatwave of 2003." *Nature* 432.7017 (2004): 610–14. http://www.nature.com/nature/journal/v432/n7017/abs/nature 03089.html.

Strong, Aaron, Sallie Chisholm, Charles Miller, and John Cullen. "Ocean Fertilization: Time to Move on." *Nature* 461.7262 (2009): 347–48. http:// www.nature.com/nature/journal/v461/n7262/full/461347a.html.

Summers, Lawrence H. "Comments on Richard Zeckhauser's Investing in the Unknown and Unknowable." *Capitalism and Society* 1.2 (September 2006). http://dx.doi.org/10.2202/1932-0213.1012.

Sunstein, Cass R. "Of Montreal and Kyoto: A Tale of Two Protocols." *Harvard Environmental Law Review* 31 (2007): 1. http://heinonline.org /HOL/LandingPage?collection=journals&handle=hein.journals/helr 31&div=5&id=&page=.

———. *Worst-Case Scenarios*. Harvard University Press, 2007. http:// www.hup.harvard.edu/catalog.php?isbn=9780674032514.

Survey Findings on Energy and the Economy. Institute for Energy Research, July 12 2013. http://www.instituteforenergyresearch.org/wp-content /uploads/2013/07/IER-National-Survey.-Key-Findings.pdf.

Taleb, Nassim Nicholas. *The Black Swan: The Impact of the Highly Improbable Fragility*. Random House, 2010. http://www.fooledbyrandomness .com.

Talke, S. A., P. Orton, and D. A. Jay. "Increasing Storm Tides in New York Harbor, 1844–2013." *Geophysical Research Letters* 41.9 (2014): 3149–55. http://onlinelibrary.wiley.com/doi/10.1002/2014GL059574/abstract.

"Technical Update of the Social Cost of Carbon for Regulatory Impact Analysis under Executive Order 12866." United States Government

Interagency Working Group on Social Cost of Carbon (November 1, 2013). http://www.whitehouse.gov/sites/default/files/omb/assets/inforeg/technical-update-social-cost-of-carbon-for-regulator-impact-analysis.pdf.

Thomas, E. "Biogeography of the Late Paleocene Benthic Foraminiferal Extinction." In *Late Paleocene–Early Eocene Climatic and Biotic Events in the Marine and Terrestrial Records*, edited by M. P. Aubry, S. C. Lucas, and W. A. Berggren. Columbia University Press, 1998. 214–43. http://books.google.com/books?hl=en&lr=&id=BR-zAQAAQBAJ&oi.

Thomson, Judith Jarvis. "The Trolley Problem." *Yale Law Journal* 94.6 (1985): 1395–415. http://www.jstor.org/stable/796133.

Thøgersen, John, and Tom Crompton. "Simple and Painless? The Limitations of Spillover in Environmental Campaigning." WWF United Kingdom (2009). http://www.wwf.org.uk/research_centre/research_centre_results.cfm?uNewsID=2728.

"Threat of Space Objects Demands International Coordination, UN Team Says." UN News Center (February 2013). http://www.un.org/apps/news/story.asp?NewsID=44186&Cr=outer+space&Cr1=#.UnAt3JQ-vvA.

Titmuss, Richard M. *The Gift Relationship.* Allen and Unwin, 1970.

"Tobacco Shares Fall on Australian Packaging Rule." *Telegraph* (August 15, 2013). http://www.telegraph.co.uk/finance/newsbysector/retailandconsumer/9476621/Tobacco-shares-fall-on-Australian-packaging-ruling.html.

Tol, Richard S. J. "Correction and Update: The Economic Effects of Climate Change." *Journal of Economic Perspectives* 28.2 (2014): 221–26. http://pubs.aeaweb.org/doi/pdfplus/10.1257/jep.28.2.221.

———. "The Economic Impact of Climate Change in the 20th and 21st Centuries." *Climatic Change* 117.4 (2013): 795–808. http://link.springer.com/article/10.1007%2Fs10584-012-0613-3.

———. "Estimates of the Damage Costs of Climate Change—Part I: Benchmark Estimates." *Environmental and Resource Economics* 21.1 (2002): 47–73. http://link.springer.com/article/10.1023%2FA%3A1014500930521.

Tollefson, Jeff. "Hurricane Sandy Spins Up Climate Discussion." *Nature: News* (October 30, 2012). http://www.nature.com/news/hurricane-sandy-spins-up-climate-discussion-1.11706.

———. "Ocean-Fertilization Project off Canada Sparks Furore." *Nature* 490.7421 (2012): 458–59. http://www.nature.com/news/ocean-fertilization-project-off-canada-sparks-furore-1.11631.

Trenberth, Kevin E., and Aiguo Dai. "Effects of Mount Pinatubo Volcanic Eruption on the Hydrological Cycle as an Analog of Geoengineering."

Geophysical Research Letters 34: L15702 (2007). http://onlinelibrary
.wiley.com/doi/10.1029/2007GL030524/abstract.

Tyndall, John. "XXIII. On the Absorption and Radiation of Heat by
Gases and Vapours, and on the Physical Connexion of Radiation,
Absorption, and Conduction—the Bakerian Lecture." *London, Edinburgh, and Dublin Philosophical Magazine and Journal of Science* 22.146
(1861): 169–94. http://www.gps.caltech.edu/~vijay/Papers/Spectro
scopy/tyndall-1861.pdf.

United Nations. *Our Common Future*. Report of the World Commission
on Environment and Development (1987). http://www.un-documents
.net/our-common-future.pdf.

U.S. Department of Defense. "Quadrennial Defense Review Report."
(February 2010). http://www.defense.gov/qdr/qdr%20as%20of%2026
jan10%200700.pdf.

U.S. Environmental Protection Agency. "Climate Change Indicators in
the United States: U.S. and Global Temperature." 2012. http://www
.epa.gov/climatechange/science/indicators/weather-climate/tempera
ture.html.

———. "Drinking Water Contaminants" (last updated June 2013).
http://water.epa.gov/drink/contaminants/.

———. "Overview of Greenhouse Gases: Carbon Dioxide Emissions."
http://www.epa.gov/climatechange/ghgemissions/gases/co2.html.

U.S. Global Change Research Program. "Climate Change Impacts in the
United States." 2014. http://nca2014.globalchange.gov/.

U.S. Supreme Court. Global-Tech Appliances, Inc., et al. v. SEB S.a. 563
No. 10–6 (May 31, 2011). http://www.supremecourt.gov/opinions
/10pdf/10-6.pdf.

van Benthem, Arthur, Kenneth Gillingham, and James Sweeney.
"Learning-by-Doing and the Optimal Solar Policy in California." *Energy Journal* 29.3 (2008): 131–52. http://ideas.repec.org/a/aen/journl
/2008v29-03-a07.html.

van den Bergh, J.C.J.M., and W.J.W. Botzen. "A Lower Bound to the Social
Cost of CO_2 Emissions." *Nature Climate Change* 4.4 (2014): 253–58.
http://www.nature.com/nclimate/journal/v4/n4/full/nclimate2135.html.

Viscusi, W. Kip, and Joseph E. Aldy. "The Value of a Statistical Life: A
Critical Review of Market Estimates throughout the World." *Journal of Risk and Uncertainty* 27.1 (2003): 5–76. http://www.nber.org
/papers/w9487.

Vivid Economics. "The Implicit Price of Carbon in the Electricity Sec-
tor of Six Major Economies." Report Prepared for the Climate Institute (October 2010). http://www.vivideconomics.com/docs/Vivid%20
Econ%20Implicit%20Carbon%20Prices.pdf.

Voosen, Paul. "Provoked Scientists Try to Explain Lag in Global Warming." Environment and Energy Publishing, 2011. http://www.eenews.net/stories/1059955427.

Wagner, Gernot. *But Will the Planet Notice? How Smart Economics Can Save the World.* Hill and Wang, 2011. http://www.gwagner.com/planet.

———. "Going Green but Getting Nowhere." *New York Times* (September 8, 2011). http://www.nytimes.com/2011/09/08/opinion/going-green-but-getting-nowhere.html.

———. "Naomi Klein Is Half Right: Distorted Markets Are the Real Problem." Grist.org (March 14, 2012). http://grist.org/climate-change/naomi-klein-is-half-right-distorted-markets-are-the-real-problem/.

Wagner, Gernot, and Martin L. Weitzman. "Playing God." Foreign Policy.com (October 24, 2012). http://www.foreignpolicy.com/articles/2012/10/22/playing_god.

Wagner, Gernot, and Richard J. Zeckhauser. "Climate Policy: Hard Problem, Soft Thinking." *Climatic Change* 110.3–4 (2012): 507–21. http://link.springer.com/article/10.1007%2Fs10584-011-0067-z.

Walter, K. M., S. A. Zimov, Jeff P. Chanton, D. Verbyla, and F. S. Chapin III. "Methane Bubbling from Siberian Thaw Lakes as a Positive Feedback to Climate Warming." *Nature* 443.7107 (2006): 71–75. http://www.nature.com/nature/journal/v443/n7107/abs/nature05040.html.

Weaver, R. Kent. "The Politics of Blame Avoidance." *Journal of Public Policy* 6.4 (1986): 371–98. http://dx.doi.org/10.1017/S0143814X00004219.

Weitzman, Martin L. "Can Negotiating a Uniform Carbon Price Help to Internalize the Global Warming Externality?" NBER Working Paper No. 19644 (November 2013), forthcoming in the *Journal of the Association of Environmental and Resource Economists.* http://www.nber.org/papers/w19644.pdf.

———. "Fat-Tailed Uncertainty in the Economics of Catastrophic Climate Change." *Review of Environmental Economics and Policy* 5.2 (2011): 275–92. http://scholar.harvard.edu/weitzman/publications/fat-tailed-uncertainty-economics-catastrophic-climate-change-0.

———. "Fat Tails and the Social Cost of Carbon." *American Economic Review: Papers and Proceedings* 104.5 (2014): 544–46. http://dx.doi.org/10.1257/aer.104.5.544.

———. "Gamma discounting." *American Economic Review* 91.1 (2001): 260–71. http://scholar.harvard.edu/files/weitzman/files/gamma_discounting.pdf.

———. "GHG Targets as Insurance against Catastrophic Climate Damages." *Journal of Public Economic Theory* 14.2 (2012): 221–44. http://scholar.harvard.edu/weitzman/publications/ghg-targets-insurance-against-catastrophic-climate-damages-0.

———. "The Geoengineered Planet." In *In 100 Years*, edited by Ignacio Palacios-Huerta. MIT Press, 2013. 145–63. http://scholar.harvard .edu/weitzman/publications/one-hundred-years-chapter-10-geo engineered-planet.

———. "On Modeling and Interpreting the Economics of Catastrophic Climate Change." *Review of Economics and Statistics* 91.1 (2009):1–19. http://scholar.harvard.edu/weitzman/publications/modeling-and -interpreting-economics-catastrophic-climate-change.

———. "Prices vs. Quantities." *Review of Economic Studies* 41.4 (1974): 477–91. http://www.jstor.org/discover/10.2307/2296698.

———. "Recombinant Growth." *Quarterly Journal of Economics* 113.2 (1998): 331–60. http://qje.oxfordjournals.org/content/113/2/331.short.

———. "A Review of the Stern Review on the Economics of Climate Change." *Journal of Economic Literature* 45.3 (2007): 703–24. http:// www.aeaweb.org/articles.php?doi=10.1257/jel.45.3.703.

———. "Subjective Expectations and Asset-Return Puzzles." *American Economic Review* 97.4 (2007): 1102–30. http://www.jstor.org /discover/10.2307/30034086.

———. "A Voting Architecture for the Governance of Free-Driver Externalities, with Application to Geoengineering." *Scandinavian Journal of Economics*, forthcoming. http://scholar.harvard.edu/weitzman/publications /voting-architecture-governance-free-driver-externalities-application.

———. "What Is the 'Damages Function' for Global Warming—and What Difference Might it Make?" *Climate Change Economics* 1.1(2010): 57–69. http://scholar.harvard.edu/weitzman/publications /what-damages-function%E2%80%9F-global-warming-%E2%80%93 -and-what-difference-might-it-make.

"What We Know: The Realities, Risks and Response to Climate Change." American Association for the Advancement of Science (2014). http://whatweknow.aaas.org/wp-content/uploads/2014/03 /AAAS-What-We-Know.pdf.

White House Press Release. August 6, 1945, Harry S. Truman Library. https://www.trumanlibrary.org/whistlestop/study_collections/bomb /large/documents/index.php?documentdate=1945-08-06&document id=59&studycollectionid=abomb&pagenumber=1.

Willis, Margaret M., and Juliet B. Schor. "Does Changing a Light Bulb Lead to Changing the World? Political Action and the Conscious Consumer." *Annals of the American Academy of Political and Social Science* 644.1 (2012): 160–90. http://ann.sagepub.com/content /644/1/160.abstract.

Wood, Graeme. "Re-engineering the Earth." *Atlantic* (June 2009). http:// www.theatlantic.com/magazine/archive/2009/07/re-engineering-the -earth/307552/.

"World Energy Outlook 2013." International Energy Agency (2013). http://www.worldenergyoutlook.org/publications/weo-2013/.

"World Energy Outlook 2014." International Energy Agency (2014). http://www.worldenergyoutlook.org/publications/weo-2014/.

"The World of Civil Aviation: Facts and Figures." International Civil Aviation Organization. http://www.icao.int/sustainability/Pages/Facts Figures.aspx.

World Resource Institute (WRI). Climate Analysis Indicators Tool. http://cait.wri.org.

Yienger, James J., Meredith Galanter, Tracey A. Holloway, Mahesh J. Phadnis, Sarath K. Guttikunda, Gregory R. Carmichael, Waller J. Moxim, and Hiram Levy II. "The Episodic Nature of Air Pollution Transport from Asia to North America." *Journal of Geophysical Research* 105.D22 (2000): 26931–45. http://onlinelibrary.wiley.com/doi /10.1029/2000JD900309/abstract.

Zeckhauser, Richard J. "Investing in the Unknown and Unknowable." *Capitalism and Society* 1.2 (2006). http://www.hks.harvard.edu/fs /rzeckhau/InvestinginUnknownandUnknowable.pdf.

Zweig, Jason. "What History Tells Us about the Market." *Wall Street Journal* (October 11, 2008). http://online.wsj.com/news/articles/SB 122368241652024977.

Index

||||||||||